Author
Wendy Weiner, M.S. Ed.
1993 Presidential Award for Science Training Excellence

Illustrators
Blanca Apodaca La Bounty
Kathy Bruce
Sue Fullam
Keith Vasconcelles

Contributing Editor
Evan D. Forbes, M.S. Ed.

Editor
Walter Kelly, M.A.

Editor-n-Chief
Sharon Coan, M.S. Ed.

Art Director
Darlene Spivak

Product Manager
Phil Garcia

Imaging
Rick Chacón

Photo Cover Credit
Images provided by PhotoDisc ©1994

Research and Contributions
Bobbie Johnson

Publishers
Rachelle Cracchiolo, M.S. Ed.
Mary Dupuy Smith, M.S. Ed.

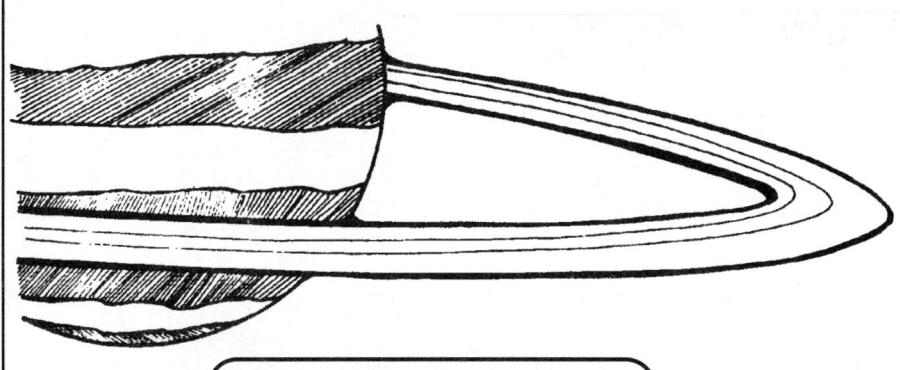

Hands-On Minds-On Science

Space
Primary

Teacher Created Materials, Inc.
6421 Industry Way
Westminster, CA 92683
www.teachercreated.com

©1994 Teacher Created Materials, Inc.
Reprinted, 2001
Made in U.S.A.
ISBN-1-55734-634-8

The classroom teacher may reproduce copies of materials in this book for classroom use only. The reproduction of any part for an entire school or school system is strictly prohibited. No part of this publication may be transmitted, stored, or recorded in any form without written permission from the publisher.

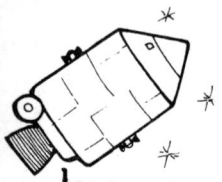

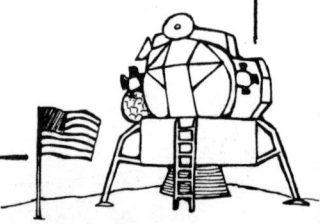

Table of Contents

Introduction 4

The Scientific Method 5

Science-Process Skills 7

Organizing Your Unit 9

What Is the History of Space Flight?

    Just the Facts 11

    Hands-On Activities

- Space Trainer 12
- 3-Stage Rocket 15
- Mercury Splashdown 17
- Hitting the Target 20
- Apollo Astronaut Egg Drop 23
- Balance the Shuttle 25
- The Shape of Things to Come 28

What Is Around and Beyond Earth?

    Just the Facts 30

    Hands-On Activities

- Trapping the Sun's Rays 32
- Space Satellite Mobile 34
- Earth and Moon 36
- Sun Yourself 38
- Our Solar System 40
- Constellation Viewer 42

#634 Space   ©1994 Teacher Created Materials, Inc.

Table of Contents (cont.)

What Would Living and Working in Space Be Like?
- Just the Facts 45
- Hands-On Activities
  - Super Space Team 47
  - My Possessions 50
  - Commmunicate It 52
  - Record Keeping 54
  - Space Station Creation 56
  - Crystal Growth 58
  - New Worlds 60

Curriculum Connections
- Language Arts 62
- Social Studies 64
- Physical Education 68
- Math 66
- Art 67
- Music 68

Station-to-Station Activities
- Mission Alpha 69
- Mission Beta 71
- Mission Gamma 73
- Mission Mu 76
- Mission Phi 78

Management Tools
- Science Safety 81
- Space Journal 82
- Space Observation Area 88
- Assessment Forms 89
- Science Awards 92

Glossary 93

Bibliography 95

©1994 Teacher Created Materials, Inc. #634 Space

Introduction

What Is Science?

What is science to a young child? Is it something that they know is a part of their world? Is it a textbook in the classroom? Is it a tadpole changing into a frog? Is it a sprouting seed, a rainy day, a boiling pot, a turning wheel, a pretty rock, or a moonlit sky? Is science fun and filled with wonder and meaning? What does science mean to children?

Science offers you and your eager children opportunities to explore the world around you and to make connections between the things you experience. The world becomes your classroom, and you, the teacher, a guide.

Science can, and should, fill children with wonder. It should cause them to be filled with questions and the desire to discover the answers to their questions. And, once they have discovered the answers, they should be actively seeking new questions to answer!

The books in this series give you and your children the opportunity to learn from the whole of your experience—the sights, sounds, smells, tastes, and touches, as well as what you read, write about, and do. This whole-science approach allows you to experience and understand your world as you explore science concepts and skills together.

What Is Space?

As long as people have been living on the Earth, they have looked into the sky with amazement and wondered. What is out there? It is curiosity that causes us to explore the unknown and learn.

The study of space started with the dawn of civilization and continues today. Chinese astronomers charted star positions and recorded eclipses of both sun and moon as early as the 1300's B.C. Then in the 1500's, Copernicus proposed the theory of a sun-centered universe, which opposed Ptolemy's theory of an earth-centered universe. This was the beginning of modern astronomy. Over the next several hundred years, new theories were developed and old theories were changed or disregarded by scientists like Galileo, Newton, and Einstein. Then in the 1950's, the Space Age began with the first flight of an artificial satellite (Sputnik). Scientists could now study space from space. Today, space explorations have become more common, and the future looks even brighter.

#634 Space ©1994 Teacher Created Materials, Inc.

The Scientific Method

The "scientific method" is one of several creative and systematic processes for proving a given question, following an observation. When the "scientific method" is used in the classroom, a basic set of guiding principles and procedures is followed in order to answer a question. However, real world science is often not as rigid as the "scientific method" would have us believe.

This systematic method of problem solving will be described in the paragraphs that follow.

1 Make an OBSERVATION.

The teacher presents a situation, gives a demonstration, or reads background material that interests students and prompts them to ask questions. Or students can make observations and generate questions on their own as they study a topic.

Example: Show actual footage of a space capsule returning to earth.

2 Select a QUESTION to investigate.

In order for students to select a question for a scientific investigation, they will have to consider the materials they have or can get, as well as the resources (books, magazines, people, etc.) actually available to them. You can help them make an inventory of their materials and resources, either individually or as a group.

Tell students that in order to successfully investigate the questions they have selected, they must be very clear about what they are asking. Discuss effective questions with your students. Depending upon their level, simplify the question or make it more specific.

Example: How did the first space capsule in the U.S. return to earth?

3 Make a PREDICTION (Hypothesis).

Explain to students that a hypothesis is a good guess about what the answer to a question will probably be. But they do not want to make just any arbitrary guess. Encourage students to predict what they think will happen and why.

In order to formulate a hypothesis, students may have to gather more information through research.

Have students practice making hypotheses with questions you give them. Tell them to pretend they have already done their research. You want them to write each hypothesis so it follows these rules:

1. It is to the point.
2. It tells what will happen, based on what the question asks.
3. It follows the subject/verb relationship of the question.

Example: I think the first space capsules in the U.S. returned to earth with the use of parachutes.

©1994 Teacher Created Materials, Inc.   5   #634 Space

The Scientific Method *(cont.)*

4 Develop a **PROCEDURE** to test the hypothesis.

The first thing students must do in developing a procedure (the test plan) is to determine the materials they will need.

They must state exactly what needs to be done in step-by-step order. If they do not place their directions in the right order, or if they leave out a step, it becomes difficult for someone else to follow their directions. A scientist never knows when other scientists will want to try the same experiment to see if they end up with the same results!

Example: You will simulate your own space capsule splashdown, with use of a parachute you create.

5 Record the **RESULTS** of the investigation in written and picture form.

The results (data collected) of a scientific investigation are usually expressed two ways—in written form and in picture form. Both are summary statements. The written form reports the results with words. The picture form (often a chart or graph) reports the results so the information can be understood at a glance.

Example: The results of your experiences can be recorded on a data-capture sheet provided (page 19).

6 State a **CONCLUSION** that tells what the results of the investigation mean.

The conclusion is a statement which tells the outcome of the investigation. It is drawn after the student has studied the results of the experiment, and it interprets the results in relation to the stated hypothesis. A conclusion statement may read something like either of the following: "The results show that the hypothesis is supported," or "The results show that the hypothesis is not supported." Then restate the hypothesis if it was supported or revise it if it was not supported.

Example: The hypothesis that stated "the first space capsules in the U.S. returned to earth with the use of parachutes," is either supported (or not supported.)

7 Record **QUESTIONS, OBSERVATIONS,** and **SUGGESTIONS** for future investigations.

Students should be encouraged to reflect on the investigations that they complete. These reflections, like those of professional scientists, may produce questions that will lead to further investigations.

Example: How do spaceships return to earth today?

Science-Process Skills

Even the youngest students blossom in their ability to make sense out of their world and succeed in scientific investigations when they learn and use the science-process skills. These are the tools that help children think and act like professional scientists.

The first five process skills on the list below are the ones that should be emphasized with young children, but all of the skills will be utilized by anyone who is truly doing science.

Observing

It is through the process of observation that all information is acquired. That makes this skill the most fundamental of all the process skills. Children have been making observations all their lives, but they need to be made aware of how they can use their senses and prior knowledge to gain as much information as possible from each experience. Teachers can develop this skill in children by asking questions and making statements that encourage precise observations.

Communicating

Humans have developed the ability to use language and symbols which allow them to communicate not only in the "here and now" but over time and space as well. The accumulation of knowledge in science, as in other fields, is due to this process skill. Even young children should be able to understand the importance of researching others' communications about science and the importance of communicating their own findings in ways that are understandable and useful to others. The space journal and the data-capture sheets used in this book are two ways to develop this skill.

Comparing

Once observation skills are heightened, students should begin to notice the relationships between things that they are observing. Comparing means noticing the similarities and differences. By asking how things are alike and different or which is smaller or larger, teachers will encourage children to develop their comparison skills.

Ordering

Other relationships that students should be encouraged to observe are the linear patterns of seriation (order along a continuum: e.g., rough to smooth, large to small, bright to dim, few to many) and sequence (order along a time line or cycle). By making graphs, time lines, cyclical and sequence drawings, and by putting many objects in order by a variety of properties, students will grow in their abilities to make precise observations about the order of nature.

Categorizing

When students group or classify objects or events according to logical rationale, they are using the process skill of categorizing. Students begin to use this skill when they group by a single property such as color. As they develop this skill they will be attending to multiple properties in order to make categorizations. The animal classification system, for example, is one system students can categorize.

Science-Process Skills (cont.)

Relating
One of the higher-level process skills, relating requires student scientists to notice how objects and phenomena interact with one another and the changes caused by these interactions. An obvious example of this is the study of chemical reactions.

Inferring
Not all phenomena are directly observable, because they are out of humankind's reach in terms of time, scale, and space. Some scientific knowledge must be logically inferred based on the data that is available. Much of the work of paleontologists, astronomers, and those studying the structure of matter is done by inference.

Applying
Even very young, budding scientists should begin to understand that people have used scientific knowledge in practical ways to change and improve the way we live. It is at this application level that science becomes meaningful for many students.

Applying
Inferring
Relating
Categorizing
Ordering
Comparing
Communicating
Observing

Organizing Your Unit

Designing a Science Lesson

In addition to the lessons presented in this unit, you will want to add lessons of your own, lessons that reflect the unique environment in which you live, as well as the interests of your students. When designing new lessons or revising old ones, try to include the following elements in your planning:

Question
Pose a question to your students that will guide them in the direction of the experience you wish to perform. Encourage all answers, but you want to lead the students towards the experience you are going to be doing. Remember, there must be an observation before there can be a question. (Refer to The Scientific Method, pages 5-6.)

Setting the Stage
Prepare your students for the lesson. Brainstorm to find out what students already know. Have children review books to discover what is already known about the subject. Invite them to share what they have learned.

Materials Needed for Each Group or Individual
List the materials each group or individual will need for the investigation. Include a data-capture sheet when appropriate.

Procedure
Make sure students know the steps to take to complete the activity. Whenever possible, ask them to determine the procedure. Make use of assigned roles in group work. Create (or have your students create) a data-capture sheet. Ask yourself, "How will my students record and report what they have discovered? Will they tally, measure, draw, or make a checklist? Will they make a graph? Will they need to preserve specimens?" Let students record results orally, using a video or audio tape recorder. For written recording, encourage students to use a variety of paper supplies such as poster board or index cards. It is also important for students to keep a journal of their investigation activities. Journals can be made of lined and unlined paper. Students can design their own covers. The pages can be stapled or be put together with brads or spiral binding.

Extensions
Continue the success of the lesson. Consider which related skills or information you can tie into the lesson, like math, language arts skills, or something being learned in social studies. Make curriculum connections frequently and involve the students in making these connections. Extend the activity, whenever possible, to home investigations.

Closure
Encourage students to think about what they have learned and how the information connects to their own lives. Prepare journals using "Space Journal" directions on page 82. Provide an ample supply of blank and lined pages for students to use as they complete the "Closure" activities. Allow time for students to record their thoughts and pictures in their journals.

Organizing Your Unit (cont.)

Structuring Student Groups for Scientific Investigations

Using cooperative learning strategies in conjunction with hands-on and discovery learning methods will benefit all of the students taking part in the investigation.

Cooperative Learning Strategies

1. In cooperative learning all group members need to work together to accomplish the task.

2. Cooperative learning groups should be heterogeneous.

3. Cooperative learning activities need to be designed so that each student contributes to the group and individual group members can be assessed on their performance.

4. Cooperative learning teams need to know the social as well as the academic objectives of a lesson.

Cooperative Learning Groups

Groups can be determined many ways for the scientific investigations in your class. Here is one way of forming groups that has proven to be successful in primary classrooms.

- **Mission Leader**—scientist in charge of reading directions and setting up equipment.
- **Specialist 1**—scientist in charge of carrying out directions (can be more than one student).
- **Specialist 2**—scientist in charge of recording all of the information.
- **Specialist 3**—scientist who translates notes and communicates findings.

If the groups remain the same for more than one investigation, require each group to vary the people chosen for each job. All group members should get a chance to try each job at least once.

Using Centers for Scientific Investigations

Set up stations for each investigation. To accommodate several groups at a time, stations may be duplicated for the same investigation. Each station should contain directions for the activity, all necessary materials (or a list of materials for investigators to gather), a list of words (a word bank) which students may need for writing and speaking about the experience, and any data-capture sheets or needed materials for recording and reporting data and findings.

Station-to-Station Activities are on pages 69-80. Model and demonstrate each of the activities for the whole group. Have directions at each station. During the modeling session, have a student read the directions aloud while the teacher carries out the activity. When all students understand what they must do, let small groups conduct the investigations at the centers. You may wish to have a few groups working at the centers while others are occupied with other activities. In this case, you will want to set up a rotation schedule so all groups have a chance to work at the centers.

Assign each team to a station, and after they complete the task described, help them rotate in a clockwise order to the other stations. If some groups finish earlier than others, be prepared with another unit-related activity to keep students focused on main concepts.

After all rotations have been made by all groups, come together as a class to discuss what was learned.

What Is the History of Space Flight?

Just the Facts

You need to know some significant things about the history of the space program.

Space flight is new to the world. It has only been since the 1960's that we have developed our space program. Prior to that people only imagined traveling into space, journeying throughout the galaxies, or walking on the moon or planets. Today space flight is a reality.

The first space flights were unmanned. They did not put people on board until they knew that a living thing could survive in space. The first living being in space was a dog. The first person was Yuri Gargarin. The first American in space was Allan Shepard, from the Mercury Missions.

At first, one person was sent into subspace. He went up and came right back down to earth. Then, we began circling the earth and returning. From there we put three men into space, during the Gemini Missions. Learning as we progressed—sometimes by trial and error, always with lots of brain power—we created better and more stable space craft.

During the Apollo Missions begun in 1969, Neil Armstrong became the first person to walk on the moon. The Mercury, Gemini, and Apollo space crafts could not be reused. These missions landed back on earth by splashing down into the ocean. They were very small and cramped.

Today we have the space transportation system. The space transportation system consists of four parts: solid rocket boosters, external fuel tank, main engines, and orbiter. We now have the technology to send eight-person teams into space and reuse not only the orbiter, but its solid rocket boosters as well. The orbiter is like a glider in the earth's atmosphere and lands like a plane. The newest orbiter, the Endeavor, replaced the Challenger, and it is even better because it can stay in space for extended periods of time and has more cargo room.

One of the most dangerous periods of space flight is during the first three minutes of liftoff. Another critical period of space travel is re-entering the earth's atmosphere. On liftoff, the astronauts deal with many factors, including the orbiter being attached to massive amounts of fuel. On re-entry, the layers of our planet's atmosphere cause friction. Objects rubbing against these gaseous layers build up heat and will burn up before they get to the ground if not protected. The orbiter must approach at the right angle and be sealed with special heat resistant tiles to protect it from burning up on re-entry. If the angle of re-entry is wrong, the shuttle could bounce off the earth's atmosphere and back into space. It might be too damaged to re-enter again.

©1994 Teacher Created Materials, Inc.   #634 Space

What Is the History of Space Flight?

Space Trainer

Question
What was done to find out if space flight was safe for people?

Setting the Stage
- Ask students what questions they might have if they did not know people could survive in space?
- Explain to students that animals and unmanned space craft were used on test runs. (The Soviets were the first and best known for these types of trials.)
- Tell students the first astronauts were dogs and monkeys.

Materials Needed for Each Group
- balloon, 9" (22.5 cm) or larger
- flexible straw
- straight pin
- pencil with an eraser
- diagram and test pilot (page 13), one per student
- transparent tape
- data-capture sheet (page 14), one per student

Procedure *(Student Instructions)*
1. Blow up the balloon to stretch it.
2. Tape balloon to the end of the straw (the end that does not flex).
3. Push the straight pin through the straw and into the eraser.
4. Make sure the straw will spin freely without touching the pencil.
5. Cut out the test pilot.
6. Attach the test pilot with a small piece of rolled tape to the flex end of the straw, before the curve.
7. Curve the straw, put your finger on the pin, and blow up the balloon.
8. Let go of the pin and put your finger over the straw after you have blown up the balloon.
9. Let go of the straw and hold onto the pencil. Observe what happens and record information on your data-capture sheet.

Extensions
- Have students repeat the experience, this time trying different balloons.
- Have students examine Newton's laws of motion.
- Discuss with students the pros and cons of animal testing.

Closure
In their space journals, have students write about their test pilot experience.

What Is the History of Space Flight?

Space Trainer (cont.)

Diagram for constructing your space trainer and your test pilot.

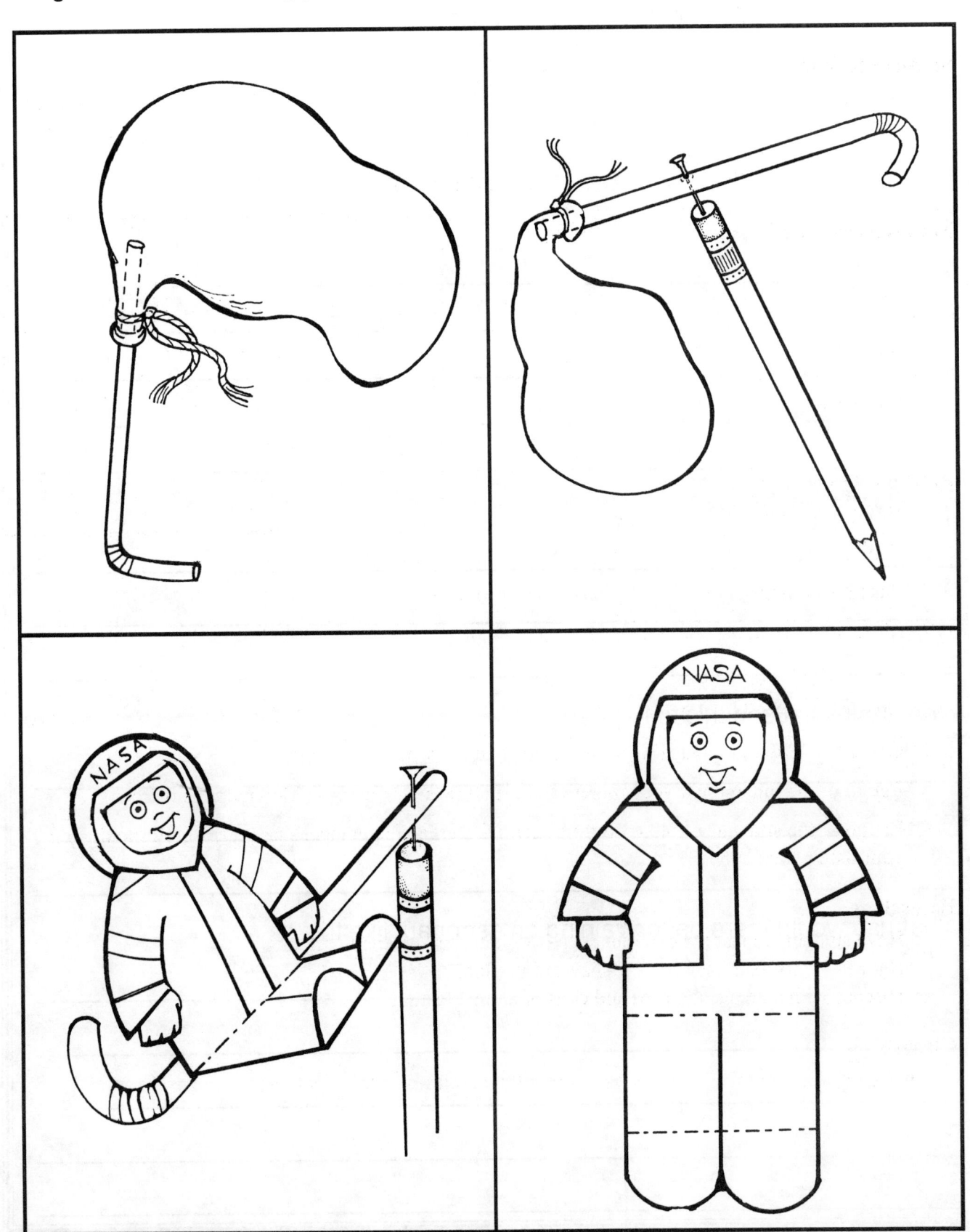

What Is the History of Space Flight?

Space Trainer *(cont.)*

Fill in the information below.

Materials I used:

_____ _____ _____

_____ _____ _____

Steps I completed:

1. _____
2. _____
3. _____
4. _____
5. _____

Results of what happened:

_____

_____

_____

What I think caused this to happen:

_____

_____

_____

What use would there be for training an astronaut like this?

_____

_____

_____

_____

What Is the History of Space Flight?

3-Stage Rocket

Question
What kind of rocket was used on liftoff?

Setting the Stage
Discuss the 3-stage rocket with your students. A large amount of fuel is needed to launch a rocket. Each one of the three stages has a set of engines and a fuel supply. As each stage uses up its fuel, it drops away, making the rocket lighter. Finally, the capsule (which has the astronauts aboard) is lifted into orbit.

Materials Needed for Each Group
- long balloon 18" (45 cm)
- round balloon 9" (22.5 cm)
- large paper or styrofoam cup
- scissors
- paper clips
- data-capture sheet (page 16), one per student

Procedure *(Student Instructions)*
1. Cut away the bottom portion of the cup.
2. Partially inflate the long balloon (#1) and place it through the top of the cup and pull it through to the bottom of the cup.
3. The neck of balloon #1 needs to be paper-clipped to keep the air from coming out.
4. Inflate the round balloon (#2) and fit the top of it into the bottom of the cup. Make sure you hold the neck of balloon #2, to keep the air from coming out. Do not cover up the paper clip from balloon #1, when fitting balloon #2, in the cup.
5. When balloon #2 is tightly in place, remove the paper clip from balloon #1 and hold the end. If the round balloon is tightly in place, the air will not come out of balloon #1 and you can let go of the neck.
6. Let go of both balloons and observe what happens. Record the information on your data-capture sheet.

Extensions
- Have students compare the shuttle and the 3-stage rocket.
- Show students a tape of an actual 3-stage rocket launch.

Closure
One of the most dangerous times for an astronaut is sitting on thousands of tons of liquid fuel before liftoff. In their space journals, have students write a paragraph about how they would feel at liftoff if they were astronauts.

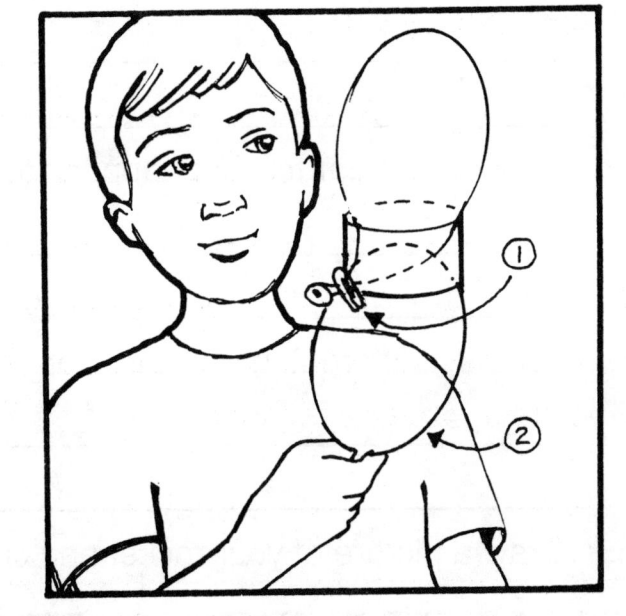

©1994 Teacher Created Materials, Inc. #634 Space

What Is the History of Space Flight?

3-Stage Rocket (cont.)

Answer the questions and draw a picture of your experience below.

1. What did you observe happening to your rocket?

   _____

   _____

2. What caused it to happen?

   _____

   _____

3. How is it similar to what happens to an actual 3-stage rocket?

   _____

   _____

4. How is it different than what happens to an actual 3-stage rocket?

   _____

   _____

5. Draw a picture of your rocket and compare it to an actual 3-stage rocket.

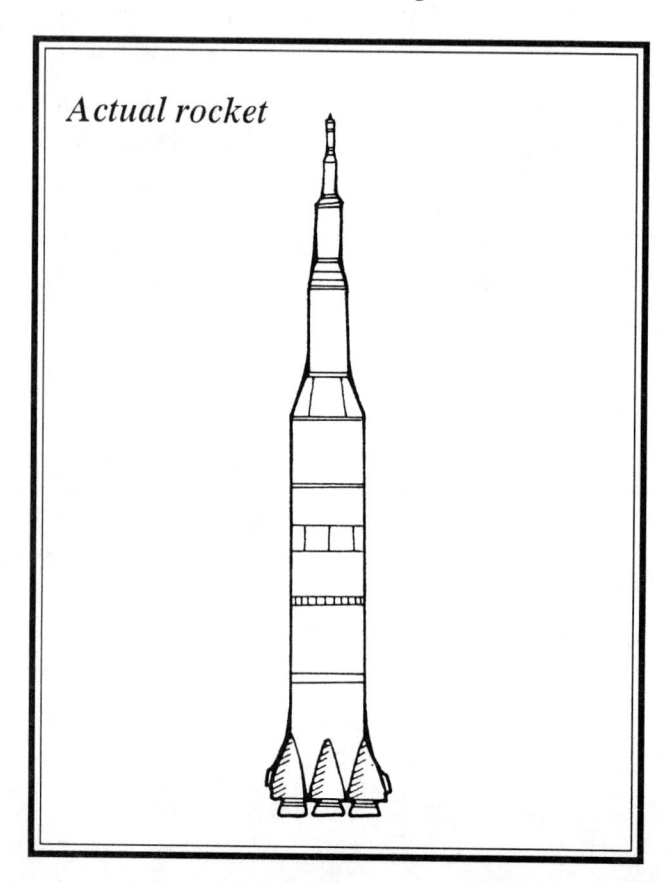

Your rocket	*Actual rocket*

#634 Space

What Is the History of Space Flight?

Mercury Splashdown

Question
How did the first space capsules in the U.S. return to earth?

Setting the Stage
- Discuss with your students how the first space capsules (Mercury, Gemini, and Apollo) did not have landing gear like today's space shuttles.
- Discuss with your students why the easiest way to bring the astronauts down was a splashdown in the ocean.
- Discuss with your students why the capsules had parachutes to slow them and cushion the impact in the water after re-entry.
- Have students define the term *splashdown*.

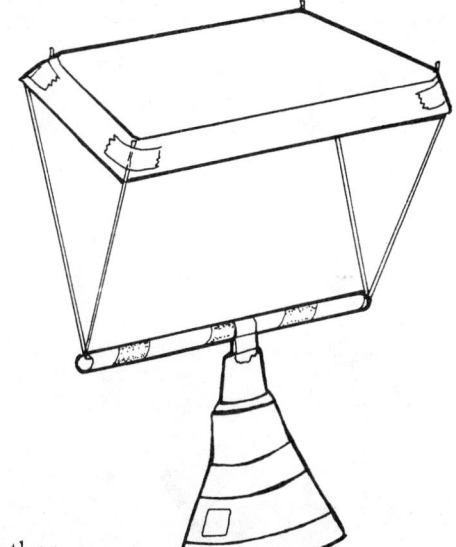

Materials Needed for Each Group
- meat tray or disposable plastic plate
- two pieces of string or yarn 24" (61 cm) long
- one straw 3" (7.5 cm) long
- reproducible capsule, trace on cardstock (page 18)
- plastic counters
- clear tape
- crayons or colored markers
- data-capture sheet (page 19), one per student

Procedure *(Student Instructions)*
1. Thread both pieces of string or yarn through the straw together.
2. Slide the straw to the middle of the string or yarn, making sure that both sides are even.
3. Tape each end of the string or yarn to each of the corners of the meat tray or equally around the plate.
4. Tape the space capsule to the middle of the straw.
5. Once you have built the parachute and attached the space capsule, it is time to try a successful splashdown.
6. Pick a spot in the classroom where students can safely reach a high place to drop their space capsules.
7. Use the plastic counters as weights if better stability is needed.
8. Record your results on the data-capture sheet.

Extensions
- Have students use different shapes and sizes of trays and plates.
- Have students use different types of materials to make parachutes.
- Have students draw an "X" on the floor and see how close they can come to landing on it.

Closure
- In their space journals, have students write about how it would feel to splashdown in the ocean.
- Draw a picture of what it might look like.

©1994 Teacher Created Materials, Inc.　　　　　　#634 Space

What Is the History of Space Flight?

Mercury Splashdown (cont.)

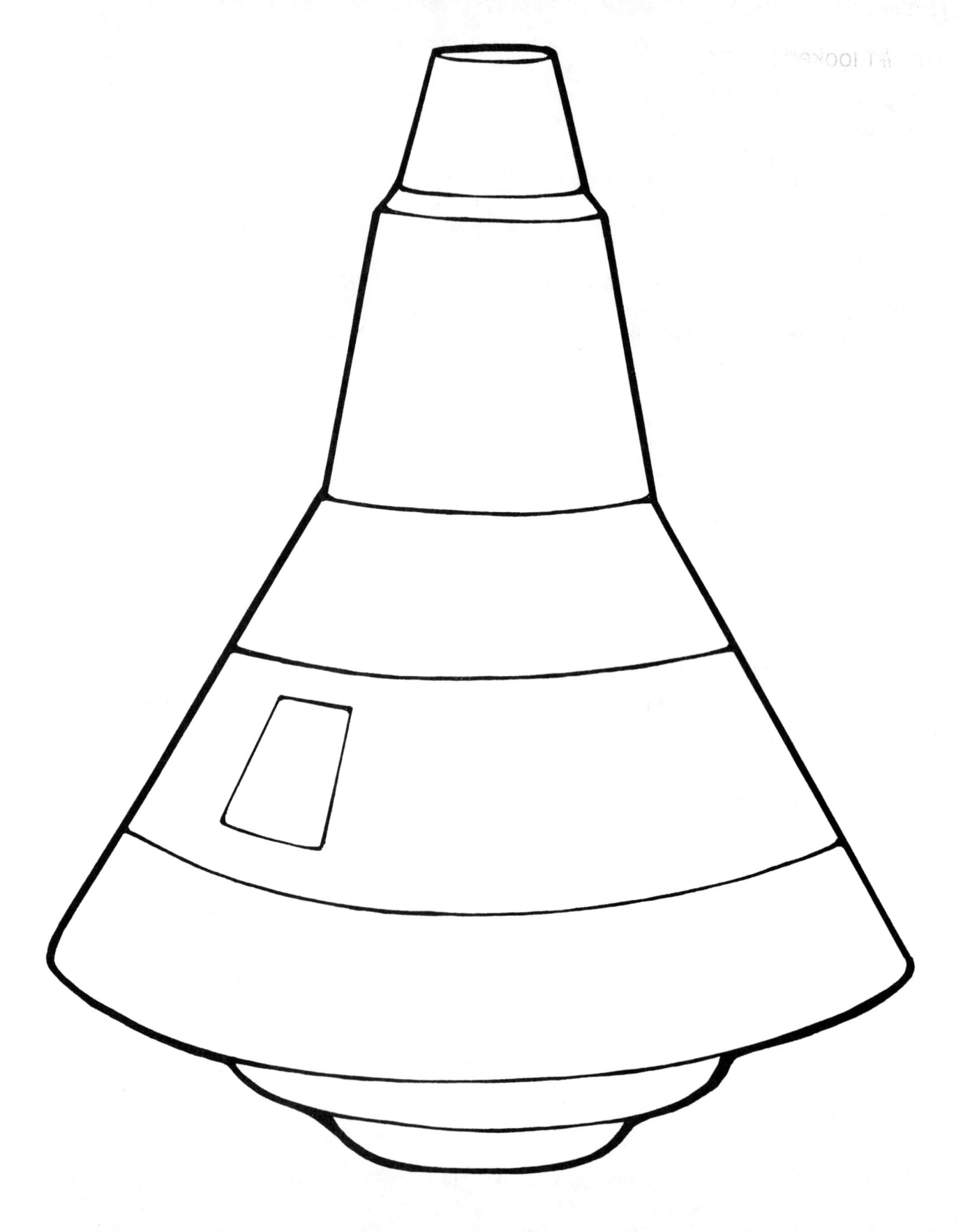

What Is the History of Space Flight?

Mercury Splashdown (cont.)

Draw a picture for each trial splashdown and answer the questions below.

Trial #1 looked like this:

How could I change the design to make trial #2 better?

What I changed: _____

Trial #2 looked liked this:

How could I change the design to make trial #3 better?

What I changed: _____

Trial #3 looked like this:

What I learned from doing all three trials: _____

_____

_____

©1994 Teacher Created Materials, Inc.   19   #634 Space

What Is the History of Space Flight?

Hitting the Target

Question

How difficult was it to retrieve the astronauts from the ocean?

Setting the Stage

- Discuss with students all the variables the Air Force was faced with when retrieving astronauts from the ocean (e.g., water currents, waves, wind currents, etc.).
- Discuss with students the difficulties involved in getting the astronauts out of the capsule, once it had landed in the ocean.

Materials Needed for Each Group

- large pan of water
- 2-liter soda bottle
- scissors
- fan
- helicopter spinner pattern (page 21), one per student
- paper clips (weights)
- data-capture sheet (page 22), one per student

Procedure *(Student Instructions)*

1. Cut soda bottle 5" (12.5 cm) from the top. Remove the solid plastic bottom. (Soak in warm water.) Reassemble the top of your bottle with the plastic bottom, and now you have your capsule.

2. Float your capsule inside the pan of water.

3. Place the fan close enough to the pan to create wind and waves when the fan is on.

4. Cut out and assemble your helicopter spinner.

5. Once your helicopter is assembled, fly it and see how close you can come to the capsule in the water. Observe what happens and record the information on your data-capture sheet.

6. You may have to add weights to adjust its flight pattern. Then repeat step #5.

Extensions

- Have students repeat the experience, this time making their helicopters out of tagboard.
- Have students adjust their fans to varying speeds.

Closure

Have students complete their data-capture sheets and add them to their space journals.

What Is the History of Space Flight?

Hitting the Target *(cont.)*

Reproducible helicopter pattern and directions for making it.

1. Use the pattern on this page to make your helicopter for hitting the target.

2. Cut out pattern along the solid blank lines.

3. Next, make cuts along the dashed lines. Fold the long strips in opposite directions. It should now look as if it has bunny ears.

4. Once the top of the helicopter is made, you need to fold the cut sides into the middle, making a stem.

5. When the stem is made, you will need to fold the bottom up 1" (2.5 cm) in order to keep the helicopter together. Test your helicopter and make it fly.

©1994 Teacher Created Materials, Inc.   #634 Space

What Is the History of Space Flight?

Hitting the Target *(cont.)*

Draw a picture of your experience and answer the questions below.

Draw your first capsule recovery:

What were some of the things that made it difficult to land near the capsule?

_____

_____

How did you improve your helicopter to make a better landing near the capsule?

_____

_____

Draw your second capsule recovery:

What Is the History of Space Flight?

The Apollo Astronaut Egg Drop

Question
How difficult is it to protect an astronaut in the space capsule?

Setting the Stage
- Discuss with your students Newton's laws of motion and how they would apply to a space capsule. See glossary for definitions.
- Discuss with your students the need to protect an astronaut from impact.

Materials Needed for Each Group
- 2-liter soda bottle
- raw egg
- various types of packaging materials (paper, cotton, styrofoam, or corn starch peanuts, plastic, etc.)
- data-capture sheet (page 24), one per student

Procedure *(Student Instructions)*
1. Using the 2-liter bottle, cut the top off 5-6" (12.5-15 cm) below the cap.
2. Then, soak the bottom of the bottle in warm water to remove the base.
3. Use the raw egg to represent your astronaut.
4. Using the packaging materials, pack your "astronauts" inside their capsules.
5. Once the "astronauts" have been properly packaged, place the top of the bottle inside the base to complete your space capsule. Your capsule is now ready for the drop test.
6. Drop each space capsule from a designated height and see which "astronauts" survive the impact.
7. Fill in the data-capture sheet as you complete your space capsule.

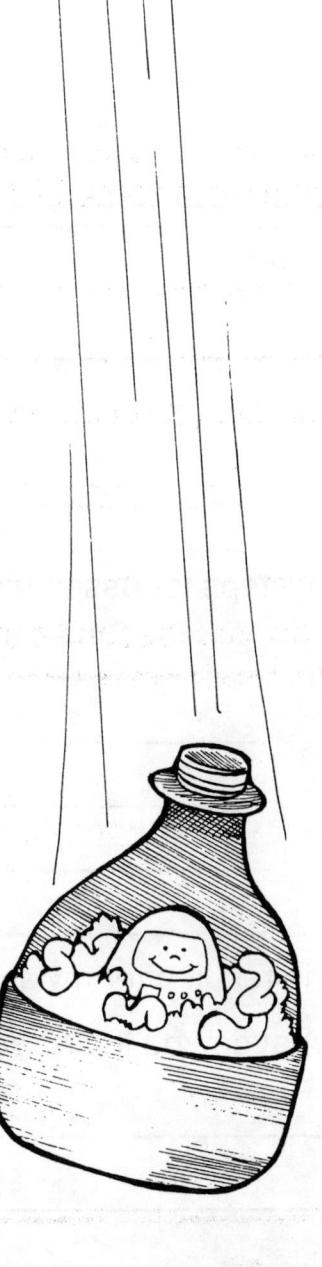

Extensions
- Have a contest with the students to see which space capsule can be dropped from the greatest height without breaking or damaging the astronaut inside.
- Have students repeat the same experience, using different containers and packaging materials.

Closure
In their space journals, have students write a story about being in a space capsule, going through re-entry, and then landing in the ocean.

©1994 Teacher Created Materials, Inc. #634 Space

What Is the History of Space Flight?

The Apollo Astronaut Egg Drop (cont.)

Draw the inside and outside of your space capsule and fill in the information needed to answer the questions below.

This is what the inside of my space capsule looks like.	This is what the outside of my space capsule looks like.

Material	Why I Used This Material

My steps for assembling my space capsule were:

1. _____
2. _____
3. _____
4. _____
5. _____

The result after my space capsule splashdown was _____ .

What do you think you would do differently next time? _____

_____

_____

What Is the History of Space Flight?

Balance the Shuttle

Question
How difficult is it to balance a space shuttle?

Setting the Stage
- Discuss with students that balance, weight, and use of space are critical factors for successful shuttle missions.
- Ask students what might happen if the shuttle is unbalanced at liftoff.

Materials Needed for Each Individual
- reproducible shuttle (page 26)
- piece of tagboard
- toothpick or small straw
- scissors
- glue
- several plastic counters (weights)
- transparent tape
- data-capture sheet (page 27)

Procedure *(Student Instructions)*
1. Cut out reproducible shuttle.
2. Glue cutout shuttle pattern to the tagboard.
3. Glue a toothpick or straw to the black line at the back center of the shuttle.
4. Try to balance the shuttle on your index finger. Observe what happens and record on your data-capture sheet.
5. Add weight to the wings where necessary to help balance the shuttle. Repeat step #4.
6. Once you have balanced the shuttle, fly it on your index finger.

Extensions
- Have students repeat the experience, this time tracing the reproducible pattern onto a piece of styrofoam. Glide when finished.
- Have students share what they learned about the importance of balance.

Closure
In their space journals, have students comment on how difficult it might be to fit all of their things into a small space.

©1994 Teacher Created Materials, Inc.   25   #634 Space

What Is the History of Space Flight?

Balance the Shuttle (cont.)

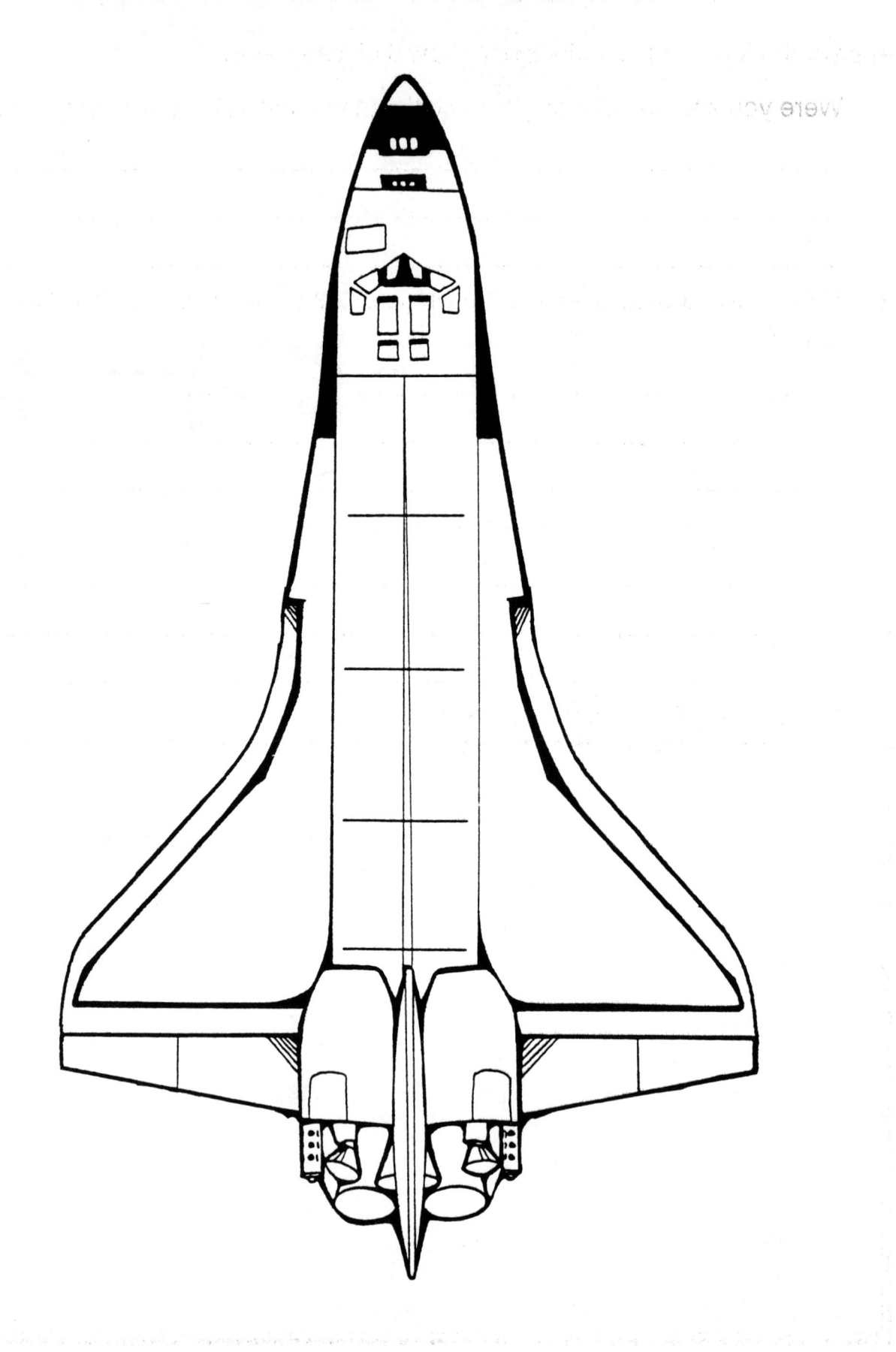

What Is the History of Space Flight?

Balance the Shuttle (cont.)

Answer the following questions and draw a picture below.

1. Were you able to balance your shuttle on the first try? Why or why not?

   _____
   _____
   _____

2. Why would adding weight to the wings of the shuttle make it easier to balance?

   _____
   _____
   _____

3. Were you successful on your second try?

   _____
   _____
   _____

4. Draw yourself flying your space shuttle.

What Is the History of Space Flight?

Shape of Things to Come

Question
What would designing a space craft be like?

Setting the Stage
- Discuss with students that with the technology available today, we could easily put a person on Mars. The problem is expense.
- Tell students that aviation-space companies are already in the design process to create an aerospace plane capable of flying in and out of earth's atmosphere.
- Share with students that a new space transportation system is also being tested to see if it is possible to use the same rockets for liftoff and landing.

Materials Needed for Each Group
- unlined paper
- drawing paper
- assorted recyclables
- glue
- transparent tape
- colored pencils or crayons
- data-capture sheet (page 29), one per student

Procedure *(Student Instructions)*
1. Discuss with your group what kind of space vehicle you would like to design.
2. Once an idea has been decided upon, you will need to make a blueprint or drawing of your space vehicle on your data-capture sheet.
3. Using your blueprint as a guide, make a drawing of your prototype on your data-capture sheet. The drawing should reflect the materials you will be using.
4. Following the drawing of your prototype, make a model of your space vehicle. If you have made any modifications to the model, draw a new picture on your data-capture sheet.

Extensions
- Have students write a letter to an aviation company, inquiring about space vehicle design.
- Have students exchange their prototypes with another group and have that group evaluate it.

Closure
In their space journals, have students write about their group's project and draw a picture of their space vehicle.

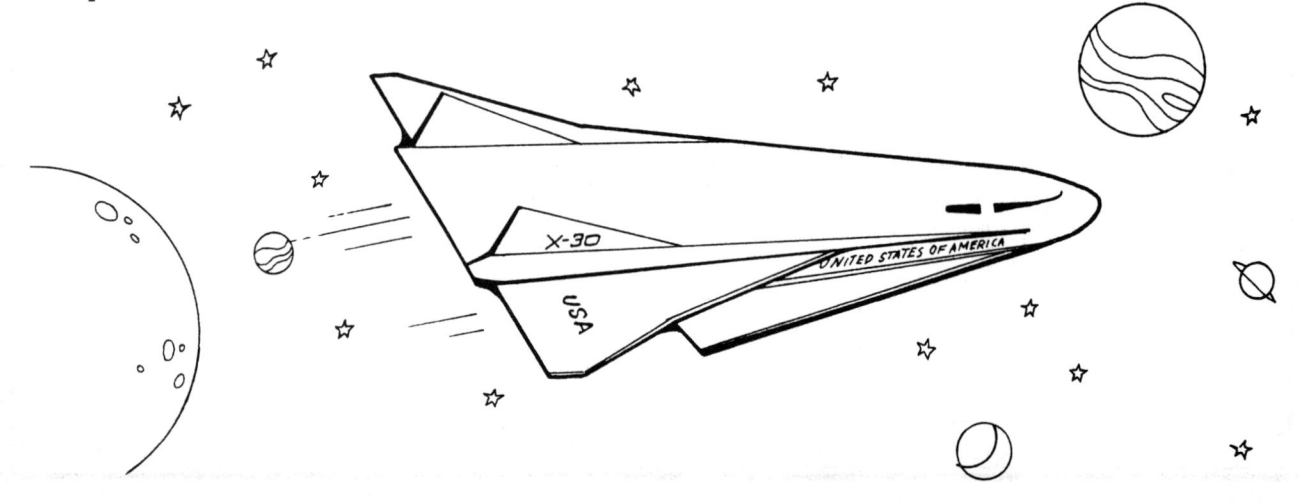

What Is the History of Space Flight?

Shape of Things to Come *(cont.)*

Complete your illustrations below.

Blueprint or Drawing

Prototype Drawing

Improvements made, if any, from Prototype Design

What is Around and Beyond Earth?

Just the Facts

Our solar system is a group of nine planets revolving around a star. The star is known as the sun. It is believed there are other solar systems in our galaxy. All things in our solar system revolve around the sun.

Our sun is a bright yellow star. Stars can be different colors depending on their intensity. The sun is 93 million miles (149 million km) from earth, but it takes only eight minutes for the sun's rays to reach the earth. The sun is very important to us. It provides warmth and light. It also enables all the plants on earth to grow. It will also be an important source of energy in the future for our planet.

The Planets

MERCURY Mercury is the planet closest to the sun. A year on Mercury lasts only 88 days, the time it takes for the planet to make one complete trip around the sun. Its surface is rocky and bare, it does not rotate on its axis, and the same side is always facing the sun. Temperatures on Mercury range between a high of 800° F (427° C), down to a low of 279° F (173° C) below 0.

VENUS Venus is the second planet from the sun and is very different from Mercury. Although Venus is rocky like Mercury, it has high mountains and deep valleys. Venus has thick yellow clouds that surround its atmosphere, making the planet surface very hot. Once the sun's heat penetrates the atmosphere, it is trapped by the clouds. This is called the Greenhouse Effect. You may have already learned about this concept. Venus can be seen from the earth and is known as the "Morning Star."

EARTH Earth is the third planet from the sun. It is the planet we live on and is the only planet at this time that we know is capable of sustaining life. The reason there is life on earth is because of its atmosphere. The planet is surrounded by a mixture of gases that allows things to live. It is the only planet with such a mixture. Seventy-percent of the planet is covered by water. The oceans make up most of the percentage, but other sources are ground water, lakes, polar ice caps, and rivers. The other thirty-percent is land in various forms–flat, mountainous, and with deep valleys such as the Grand Canyon. Because this is the only planet known to sustain life, it is important that the people that live on it, take care of it. At this time, there is a worldwide effort called Mission to the Planet Earth. With the use of satellites and shuttle missions, this effort is examining some of the environmental problems on earth, trying to figure out the answers needed to solve

What is Around and Beyond Earth?

Just the Facts *(cont.)*

MARS   Mars is the fourth planet from the sun. It is reddish in color because of iron oxidizing on the planet, much in the same way a car rusts. In order for oxidation to occur, there must be water. On the surface of Mars it is possible to see where large oceans may have existed millions of years ago. At this time, however, scientists believe there is water below the planet surface and locked in its polar caps. The atmosphere around Mars has a small amount of oxygen, and it is believed that life may exist on the planet surface. It just has not yet been found.

JUPITER   Jupiter, the fifth planet from the sun, is the largest planet in the solar system. It is estimated that 11 earths could fit across Jupiter. Jupiter has a thick atmosphere of swirling gases and seems to be in a constant storm. The large red spot seen on Jupiter is an atmospheric storm, much like a hurricane on earth, and it changes position from year to year. To date we know of 18 moons around the planet Jupiter.

SATURN   Saturn is the sixth planet from the sun. It is a planet of gas, although scientists believe the planet core is solid. The rings of Saturn are made up of ice and rocks and can be seen from earth with a telescope. Currently there are 23 moons known to be in Saturn's orbit. Titan is one of the only moons in the solar system known to have an atmosphere.

URANUS   Uranus is the seventh planet from the sun. It is a planet of gas, possibly having a solid core several thousand miles in radius. Uranus has 15 known moons and 11 thin rings around it.

NEPTUNE   Neptune is the eighth planet from the sun. It is a planet of gas. Within its atmosphere exists a dark area called the Great Dark Spot. This is a violent storm of gases, similar to the red spot on Jupiter. Six moons and several rings exist around the planet.

PLUTO   Pluto is the farthest planet from the sun. This planet, like Neptune, cannot be seen from earth without a telescope. Pluto is the only planet in the solar system that goes within the orbit of another planet. Little is known about Pluto, except that is probably one of the coldest places in our solar system.

What is Around and Beyond Earth?

Trapping the Sun's Rays

Question
How does the earth's atmosphere differ from that of other planets?

Setting the Stage
- Discuss with students the Greenhouse Effect. (A planet's atmosphere is a big factor in controlling that planet's temperature. It traps the sun's rays and determines how hot the surface of the planet will become. Some planets have thick atmospheres that keep the planets at extreme temperatures. The earth's balanced atmosphere enables life to flourish and grow. As we destroy or remove layers of the atmosphere, the earth's temperature will change.)
- Have students research how different planets are affected by their atmospheres.

Materials Needed for Each Group
- four shoe or other small boxes
- four thermometers
- tape
- plastic wrap
- data-capture sheet (page 33), one per student

Procedure (Student Instructions)
1. Remove all lids from boxes.
2. Label boxes from 1-4.
3. Place one thermometer inside each box, facing up.
4. Box 1 does not get covered. Box 2, cover with one piece of plastic wrap. Box 3, cover with two pieces of plastic wrap. Box 4, cover with three pieces of plastic wrap.
5. Tape the boxes with plastic wrap so that there is a tight seal and nothing can get inside the box.
6. Place all four boxes together in the sun.
7. Observe and record the temperatures for three days on your data-capture sheet.

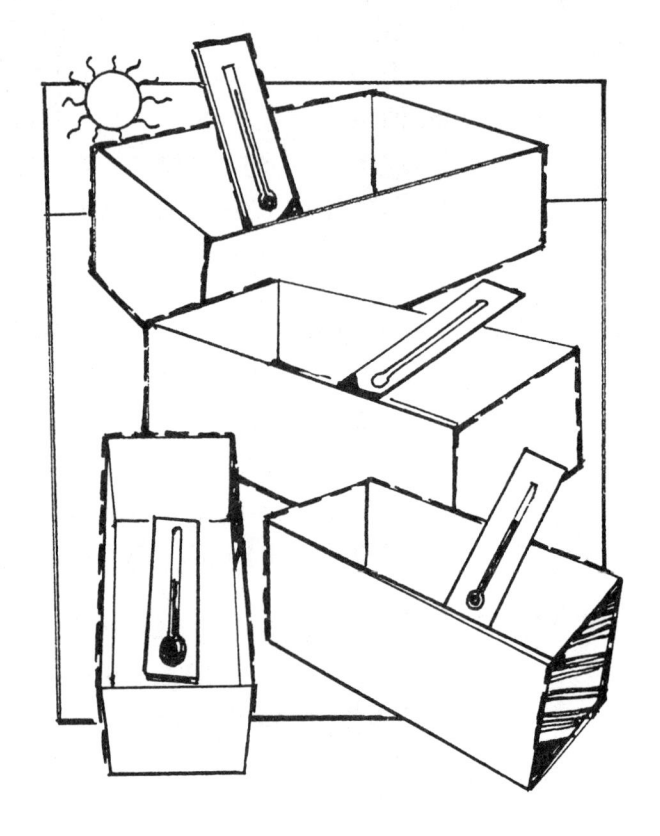

Extensions
- Have students repeat the experience, this time using different colored plastic wrap.
- Have students repeat the experience, this time painting the bottom of the boxes different colors.
- Have students repeat the experience, this time leaving space between the plastic wrap layers.

Closure
In their space journals, have students write about what they think would happen to the earth if there were no atmosphere.

What is Around and Beyond Earth?

Trapping the Sun's Rays *(cont.)*

Complete the chart below.

	Day 1	Day 2	Day 3
Temperature of Box 1			
Temperature of Box 2			
Temperature of Box 3			
Temperature of Box 4			

Findings or conclusions of experience: _____

_____

_____

_____

What is Around and Beyond Earth?

Space Satellite Mobile

Question
What kinds of things are in orbit around the earth?

Setting the Stage
- Explain to students that there are a wide variety of things that people have left in space.
- Question students to see if they know what will happen to all the "junk" left in space.

Materials Needed for Each Group
- NASA's Automated Exploration Vehicles (page 35), one per student
- tagboard
- tennis ball
- dowel rod 1" (2.5 cm) long
- blue paper
- crayons or colored markers
- clothes hanger
- string
- transparent tape
- scissors
- glue stick
- hole punch
- paper clips

Procedure (Student Instructions)
1. Unravel clothes hanger, bend it into a circle, and tape it so it stays together.
2. Cut four pieces of string the diameter of the hanger. Then, attach the four strings across the circle. Set aside until later.
3. Cover the tennis ball with blue paper and using your crayons or colored markers, make it look like the earth.
4. Cut a small hole in the tennis ball, big enough to fit the 1" (2.5 cm) dowel rod. Tie a piece of string 3" (7.5 cm) long to the center of the dowel and push it into the hole in the tennis ball.
5. Hang the earth from the center of the hanger. Set aside until later.
6. Cut out 5-6 Automated Exploration Vehicles. Glue the cut-outs to your tagboard and then cut out the vehicles around the tagboard.
7. Punch a hole in the top of each cut-out vehicle and attach a piece of string 2-4" (5-10 cm) long to each one. Make each piece of string a different length and tie the vehicles around the hanger.
8. Cut a piece of string 12" (30 cm) long. Tie a paper clip hook to one end of the string and tie the other end to the center strings of the hanger. It is now ready to hang.

Extensions
Have students research the types of things that have been left in space and report to the class.

Closure
In their space journals, have students write about how they balanced their mobiles.

What is Around and Beyond Earth?
Space Satellite Mobile *(cont.)*

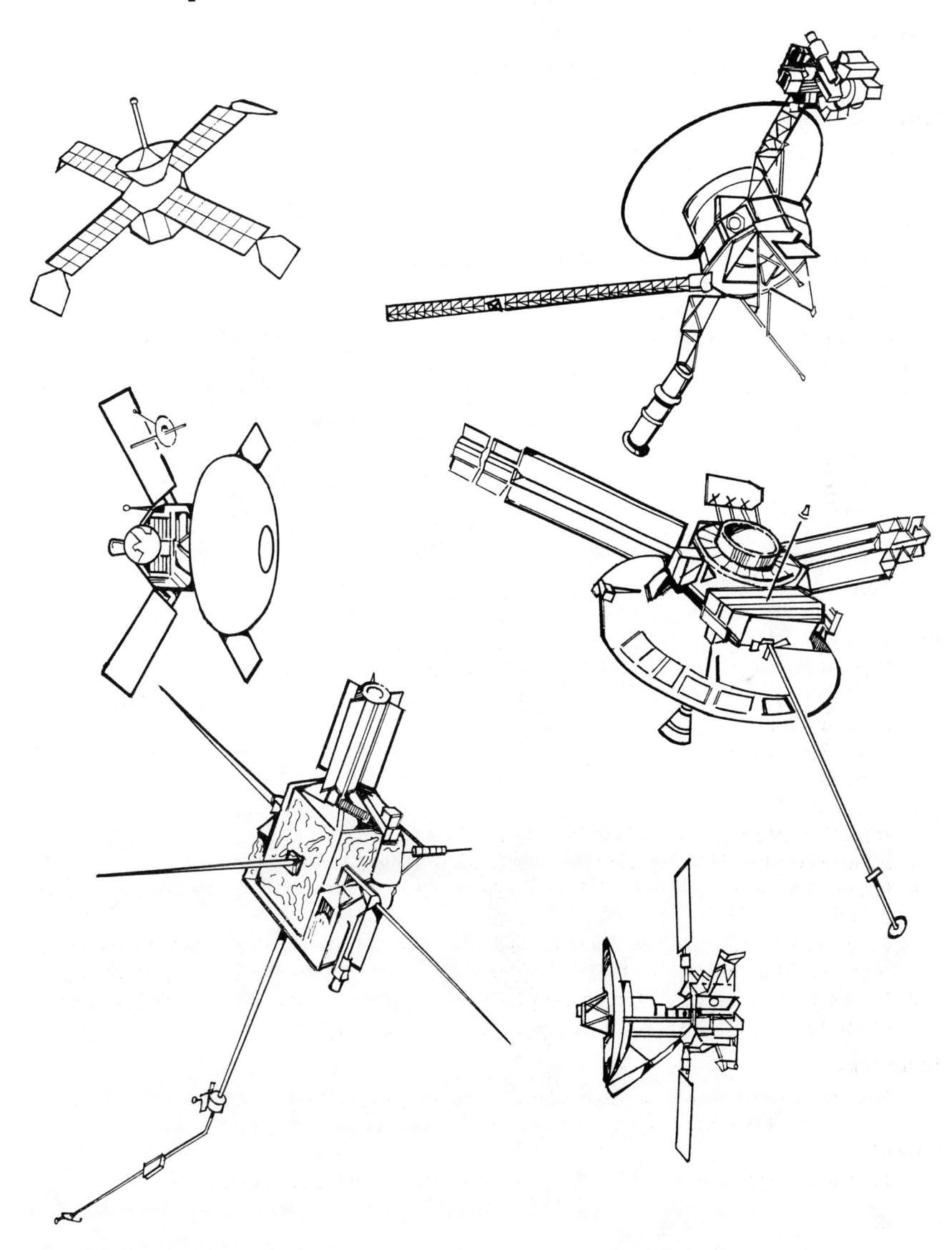

What is Around and Beyond Earth?

Earth and Moon

Question
What is the distance between the earth and the moon?

Setting the Stage
- Have students define the terms *circumference*, *diameter*, and *radius*.
- Discuss with students the concept of distance. Give examples of distances they know and then give examples of distances they do not know. Distance around the earth is 25,000 miles (40,000 km).
- Have students measure the distances between several different cities, states, and countries.

Materials Needed for Each Group
- world globe 12" (30 cm) in diameter
- tennis ball
- aluminum foil
- string or twine 20' (6-7 m) in length
- calculator
- ruler
- data-capture sheet (page 37), one per student
  Note: the measurements in this activity have been designed for the above materials. If using substitute materials, distances will have to be amended for the project to work.

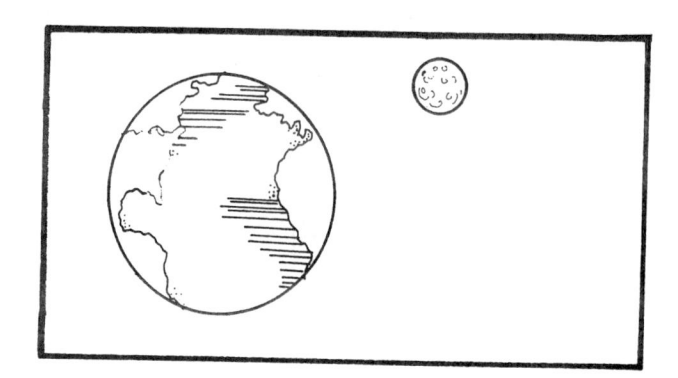

Procedure (Student Instructions)
1. Look up the distance between the earth and moon.
2. Using a calculator, divide the distance between the earth and moon by the distance around the earth. The answer to this problem will tell you how many trips around the earth represent the distance between the earth and moon.
3. Cover tennis ball with aluminum foil.
4. Wrap string or twine (x) amount of times around the world globe.
5. Unravel string or twine from the globe and measure its distance with a ruler. Each foot of string or twine measured represents part of the distance between the earth and moon. 1' (30 cm) = 25,000 miles (40,000 km).
6. Have someone in your group hold one end of the string or twine to the world globe. Have someone else stretch out the string or twine to the other side of the room. Have a third person hold the moon where the string or twine ends. This is a scale model, representing the distance between the earth and moon.

Extensions
- Have students go outside and create a larger model of the earth-moon distance.
- Have the entire class go outside and create a scale model of the solar system.

Closure
In their space journals, have students write a paragraph about being an astronaut and having to travel a great distance from home.

#634 Space   ©1994 Teacher Created Materials, Inc.

What is Around and Beyond Earth?

Earth and Moon *(cont.)*

Fill in the information needed below.

1. What is the distance in miles (kilometers) around the earth?

2. What is the longest distance you have ever travelled?

3. How many trips around the earth would you have to make, in order to make one trip to the moon?

4. Do you think in the future that flights to the moon will be as common as flying from New York City to Los Angeles? Why?

5. Draw a picture of your scale model.

What is Around and Beyond Earth?

Sun Yourself

Question
If the sun's energy can burn people, can it also cook food?

Setting the Stage
- Have your students define the term *star*.
- Discuss with your students that the closest star to the earth is the sun and it not only supplies the earth with heat energy, but being exposed to the sun for too long can be dangerous.
- Discuss with your students what solar energy is and how people use it.

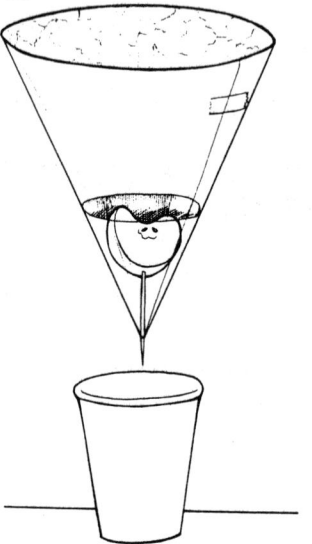

Materials Needed for Each Group
- a piece of tagboard 12" x 18" (30 cm x 45 cm) long
- aluminum foil
- black construction paper
- one small hot dog or apple
- plastic wrap
- scissors
- glue
- long toothpick
- large paper cup
- data-capture sheet (page 39), one per student

Procedure *(Student Instructions)*
1. Fit the aluminum foil onto the tagboard and glue it down as smoothly as possible.
2. When the foil has dried onto the tagboard, cut it into a circle.
3. From a piece of black construction paper, cut out a 2-3" (5-7.5 cm) circle and glue it to the middle of the foil circle. Allow time for drying.
4. Cut one line from the outside of the foil circle to the center and make it into a cone.
5. Push the toothpick through the center of the cone, attach the hot dog or apple to the top of the toothpick, and cover the cone with plastic wrap.
6. Place the cone inside the paper cup to stabilize it and then put the cooker in the sun until your food is ready to eat.
7. Enjoy what you made in your cooker, but before you eat it, complete the data-capture sheet.

Extensions
- Explain to your students why they used foil and black construction paper in the design of their solar cookers.
- Have your students design a different type of solar cooker, using different materials.

Closure
In their space journals, have your students draw a picture of their solar cookers and write about how their food tasted.

What is Around and Beyond Earth?

Sun Yourself (cont.)

Complete the questions below.

Describe what you made.

_____
_____
_____
_____

Knowing what you do about solar cookers, write four questions using that information.

1. _____
2. _____
3. _____
4. _____

Write three questions about solar cookers to which you do not know the answers.

1. _____
2. _____
3. _____

Share two of your seven questions with someone else.

1. _____
2. _____

What else would you like to know about the sun?

_____
_____
_____
_____

©1994 Teacher Created Materials, Inc.   39   #634 Space

What is Around and Beyond Earth?

Our Solar System

Question

What are the names and order of the planets in our solar system?

Setting the Stage

- Have students define the term *planet*.
- Have a class discussion about the planet in which we live.
- Have students research one of the nine planets in our solar system.

Materials Needed for Each Group

- heavy cardboard 6" x 2' (15 cm x 60 cm), to be used as a base
- 12 large paper clips
- different colored modeling clay, to represent all of the planets and the sun
- colored pictures of all the planets and the sun
- transparent tape
- 12 adhesive labels
- data-capture sheet (page 41), one per student

Procedure *(Student Instructions)*

1. Make a label for each planet in the solar system, as well as the sun.
2. Open up each paper clip so it looks like an "L."
3. To make your planet stands, tape ten of the open paper clips down on the cardboard base. They should be taped every 2.5" (6.25 cm), and the larger part of the paper clip should be sticking in the air.
4. Using the pictures of the planets and the colored modeling clay, make each planet and the sun.
5. Stick each of your planet balls and the sun on the paper clip stands, in the correct order of the solar system.
6. Stick each label in front of its correct planet.
7. After you have completed your solar system model, complete the data-capture sheet provided.

Extensions

- Have students give oral reports on their planet research.
- Take students on a field trip to a planetarium.

Closure

In their space journals, have students draw pictures of the solar system. In it they can add whatever they want (e.g., asteroids, satellites, shuttle, etc.).

What is Around and Beyond Earth?

Our Solar System (cont.)

Using a resource book, fill in the information needed below.

Planet Data:

Planets	Distance from the sun	Distance from the earth	Length of year	Temperature (high and low)	Number of satellites (moons)
Mercury					
Venus					
Earth					
Mars					
Jupiter					
Saturn					
Uranus					
Neptune					
Pluto					

What is Around and Beyond Earth?

Constellation Viewer

Question
What is a constellation?

Setting the Stage
- Have your students define the term *constellation*.
- Discuss how constellations were given names by ancient cultures to help explain how and why stars came to be in the sky. Have your students list as many constellations as they know and then create a master class list.

Materials Needed for Each Group
- a cardboard packaging box
- scissors
- a paper towel tube
- black poster paint
- aluminum foil
- black cotton thread
- clear tape
- constellation pattern (page 43), one per student
- data-capture sheet (page 44), one per student

Procedure (Student Instructions)
1. Cut off the end flaps of the box.
2. Paint the entire inside of the box, including remaining flaps, with the black poster paint.
3. Cut a hole at one end of the box and put the paper towel tube in about 1" (2.5 cm), making a viewer.
4. Using the black cotton thread, cut several lengths at various sizes. Make a knot at one end of each thread, and then wrap some foil around each knot, making a ball. These can also be various sizes.
5. Tape the strings to the top inside of the box, so the foil balls hang down in a pattern. (Example patterns will be provided on page 43.)
6. Loosely tape the large flaps so a little light is able to enter the box. Look at your constellations through the paper-towel tube viewer.
7. Complete the data-capture sheet provided.

Extensions
- Have your students share their constellations with others in the class.
- Have your students research any stories that exist about their constellations. There may be more than one story for each constellation.

Closure
In their space journals, have students create their own constellations, draw them, and then make up stories about them.

What Is Around and Beyond Earth

Constellation Viewer *(cont.)*

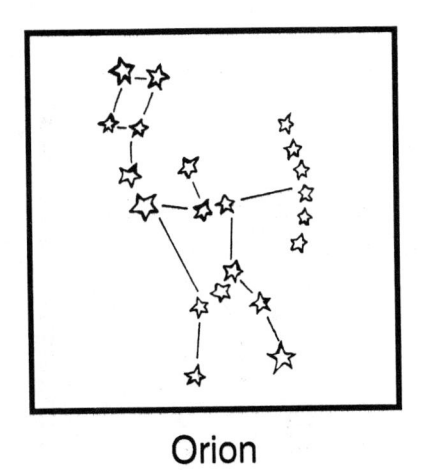

Orion	Big Dipper	Cassiopeia
Leo	Little Dipper	Andromeda
Draco	North Star	Cygnus

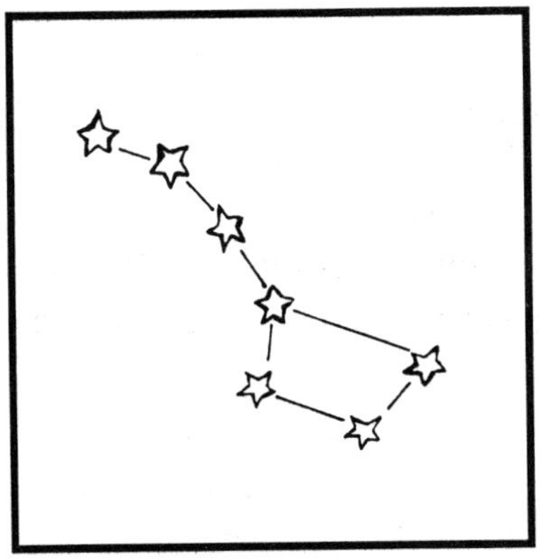

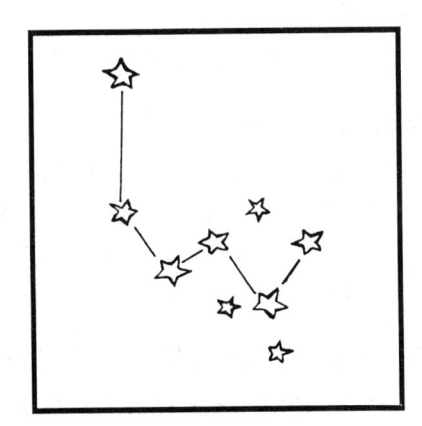

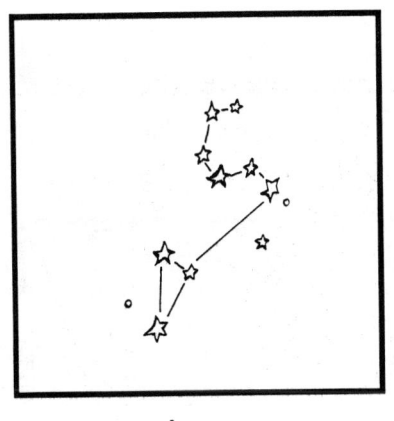

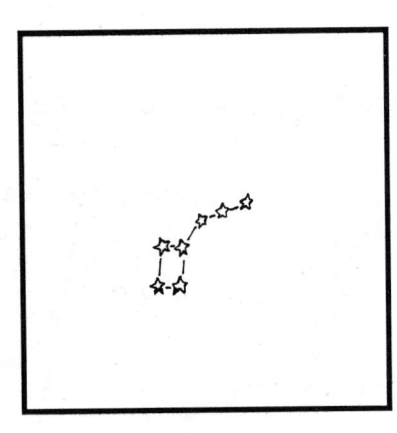

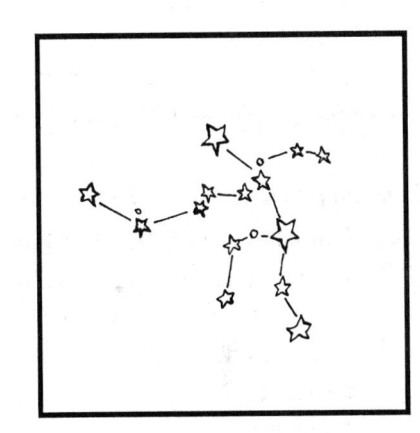

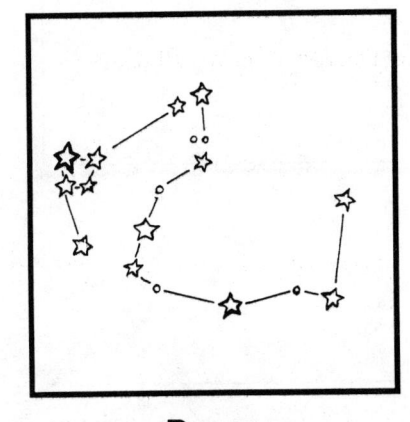

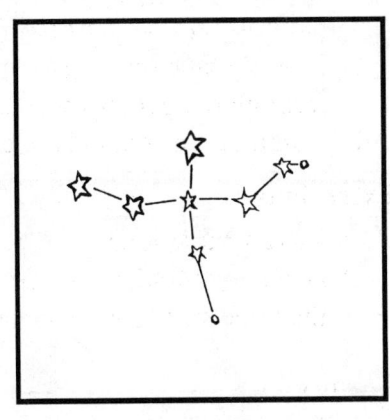

©1994 Teacher Created Materials, Inc. 43 #634 Space

What Is Around and Beyond Earth?

Constellation Viewer (cont.)

Draw and label your constellation below, just using stars.

[ ]

Draw your constellation using stars and then connect the stars to make the picture of your constellation. Label your constellation with its correct name.

[ ]

Where is your constellation located in the night sky?

_____
_____
_____

What Would Living and Working in Space Be Like?

Just the Facts

Have you ever dreamed about what it would be like to explore space? Space is not really that far. If you could drive a car straight up to space, it would take you approximately two hours to get there. Imagine yourself strapped into a space craft awaiting liftoff. The rockets ignite, and you feel your body being pushed against your seat with great force. This force is called G force. It makes you feel much, much heavier than you really are. The G force settles as soon as you are in orbit, and then unexpectedly you feel weightless. This is called zero gravity. Weightlessness influences everything in space—your health, transportation, what you eat, drink, and how you sleep.

Half of all astronauts suffer "space sickness." Your body is not used to weightlessness, and this may cause nausea and dizziness. However, after a few days, your body readjusts to this strange feeling, and the sickness goes away. Weightlessness can cause even more serious problems to your health. Without the gravity we have on earth, your muscles can become weak and may start to deteriorate. To prevent this, most spacecraft missions provide some type of exercise program. Astronauts must have physical and medical checks before and after every space visit.

In space, astronauts spend the majority of the time inside the spacecraft. However, sometimes the astronauts do need to leave the spacecraft to go outside. Leaving the spacecraft is called extravehicular activity, more commonly known as spacewalking. An astronaut needs to be protected if he or she is going to leave the spacecraft. The protection he or she uses is simply called a spacesuit. This special suit is made of many layers of fiber. The suit provides oxygen for breathing, a radio unit for communication, water for cooling purposes, and a protective layer of fiber to guard against the sun and flying particles.

The food in space is also somewhat special. The different forms of food taken on the spacecraft are dehydrated (the water is taken out of them). They are then packed in thermostabilized cans and sealed in pouches. By adding cold or hot water, the astronauts can eat foods like chicken, potatoes, beans, and even pizza. Special ovens and microwaves can be used to prepare these foods on board. Foods like nuts, cookies, and canned fruits are ready to eat "as is." The astronauts have to use special food trays. These trays are velcroed® on the bottom to stick to the astronauts' pants or shorts so they can eat without chasing their food.

What Would Living and Working in Space Be Like?

Just the Facts *(cont.)*

Sleeping, as well as eating, in space is different. Astronauts are strapped into cabins with vertical sleeping bags. Without gravity, it does not matter in which direction you sleep.

How do you think astronauts bathe? Without gravity, water does not fall to the floor in the shower. Water has to be pulled down the drain with a suction fan because it floats in a space environment. Astronauts usually vacuum excess water off their bodies before they get out of the shower. All water is carefully monitored, for any moisture getting on the delicate instruments in the spacecraft can be dangerous.

Your dream of exploring space may one day come true. In the past 25 years, space exploration has been done up close. Spacecraft have been to Mars and Venus, and astronauts have even walked on the moon. In the next century when bases have been set up on the Moon and Mars, more and more people will have a chance to explore this wonderful environment we call space.

What Would Living and Working in Space Be Like?

Super Space Team

Question
Is it hard to work on a team in an assigned position?

Setting the Stage
- Discuss with your students that each astronaut is assigned a position before a space mission takes place.
- Discuss with your students that all astronauts practice their duties over and over again before the actual mission begins.
- Discuss with your students that teamwork is very important if the mission is going to be a success.

Materials Needed for Each Group
- reproducible badges (page 48)
- seven, 3" (7.5 cm) iron nails
- hammer
- a 3" x 5" (7.5 cm x 12.5 cm) block of wood
- data-capture sheet (page 49), one per student

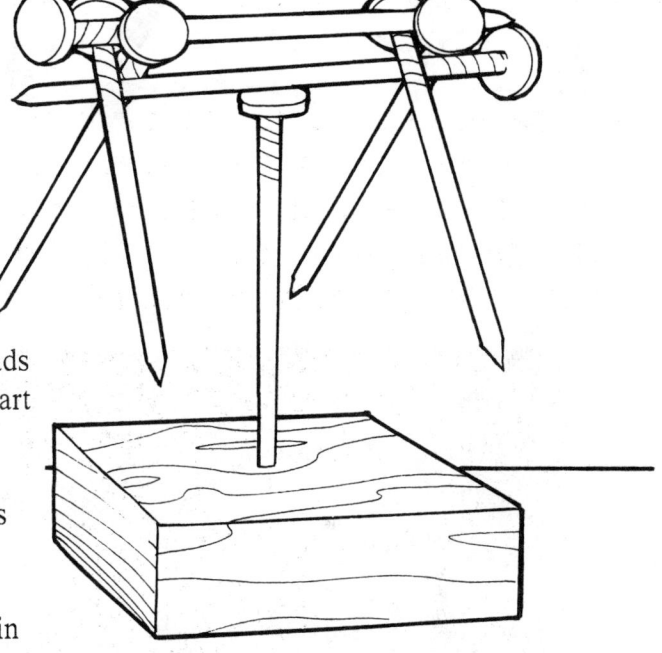

Procedure *(Student Instructions)*

1. Each student in the group needs to be assigned a position.

    Mission Leader- in charge of the project, reads the instructions and takes part in the experience.

    Specialist 1- in charge of getting all materials needed, also takes part in the experience.

    Specialist 2- in charge of cleaning up afterwards, also takes part in the experience.

    Specialist 3- in charge of keeping notes, also takes part in the experience

2. Give all students a badge to identify themselves.
3. Start out by having one student hammer one of the nails into the block of wood.
4. The assignment is to balance the remaining six nails on top of the nail that has been hammered into the block of wood.
5. Have the students work together to try to solve the problem. The idea of this experience is not necessarily to come up with the correct answer, but to work together as a team. The solution can be given at the end of the experience.
6. Answer the questions on the data-capture sheet.

Extensions
Have your students research actual space missions to see what kind of team experiments have been done.

Closure
In their space journals, have students write about the position they would most like to have if they were going to fly in the space shuttle.

©1994 Teacher Created Materials, Inc.　　　　47　　　　#634 Space

What Would Living and Working in Space be Like?

Super Space Team (cont.)

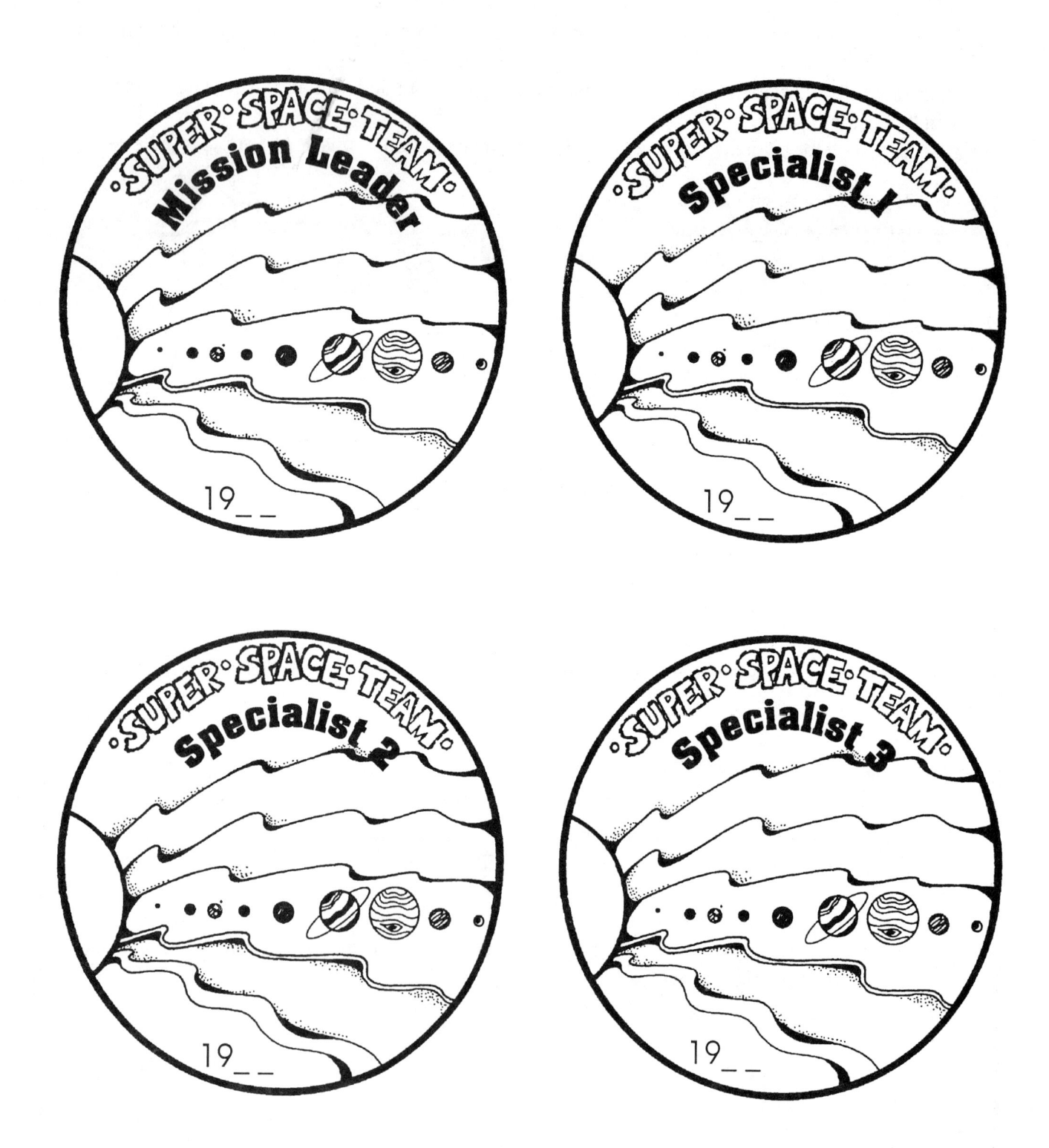

#634 Space 48 ©1994 Teacher Created Materials, Inc.

What Would Living and Working in Space Be Like?

Super Space Team *(cont.)*

Fill in the information below.

What is your assigned position on the space team? _____

What was your job in your position? _____
_____

Did you like the position you had? Why or why not? _____
_____
_____

How well did your team work together? _____
_____
_____

What might you have done differently to be a better team? _____
_____
_____

Draw a picture of your team working on your project.

What Would Living and Working in Space Be Like?

My Possessions

Question

What kind of personal items do astronauts take into space?

Setting the Stage

- Ask students, if they have ever gone away to camp. Were they able to take whatever they wanted with them, or were they limited to what they could carry?
- Ask students if they have ever gone on a trip and were allowed to take everything they wanted.
- Discuss with students that astronauts have a limited amount of space for personal items when going on a shuttle mission.

Materials Needed for Each Group

- one box approximately 14" x 14" x 14" (35 cm x 35 cm x 35 cm), one per class
- butcher or chart paper and marker, one per class
- pen or pencil
- data-capture sheet (page 51), one per student

Procedure *(Teacher/Student Instructions)*

1. Have a class discussion about allowable weight limits on the space shuttle. Have students share their ideas with the class about personal items that might be important to bring.
2. Show the box to the class and tell them this would be the approximate size of their personal locker on the space shuttle.
3. Decide what personal items you feel you would need to have on the space shuttle. Record those items on your data-capture sheet.
4. Then, after everyone in your group has completed their lists, see what items are similar, what items are different, and discuss the importance of each one.
5. Have the class come together and discuss the items each group has chosen for their personal lockers.
6. Finally, on the butcher or chart paper, make a class list of the items that the entire class decides to be most important.

Extensions

- Have students bring in the items on the class list to see if they would all fit in a personal locker.
- Discuss with students what it might be like for a new person to come to America. Often they bring only what they can carry.
- If possible, have a guest speaker come to class and talk about having to give up personal possessions.

Closure

In their space journals, have students draw a picture of their personal lockers with the items they want to take.

What Would Living and Working in Space Be Like?

My Possessions *(cont.)*

Make a list of personal possessions you are going to take on the space shuttle.

1. _____

2. _____

3. _____

4. _____

5. _____

6. _____

7. _____

8. _____

9. _____

10. _____

What Would Living and Working in Space Be Like?

Communicate It

Question
How difficult is it for Mission Control to give instructions to astronauts already in space?

Setting the Stage
- Discuss with students the importance of communication. Give examples of what happens when communication is poor. Usually the biggest problem with poor communication is a misunderstanding.
- Discuss with students the importance of following instructions when given. Give examples of what happens when instructions are not followed. Ask students what happens when you are making chocolate chip cookies and you forget the chocolate chips.

Materials Needed for Each Group
- several building blocks or anything that can be used to make a structure
- large book or piece of cardboard
- data-capture sheet (page 53), one per student

Procedure *(Student Instructions)*
1. Set a large book or piece of cardboard between you and your partner, so as to screen what each person is doing.
2. One person builds a structure.
3. Once the structure is completed, the builder needs to communicate instructions to create the same structure on the other side of the screen.
4. When both structures are complete, remove the screen to see how the structures compare. Record your observations on your data-capture sheet.
5. Repeat the activity again, but this time the person who built a structure first, will wait for instructions. Then repeat steps three and four.

Extensions
- Have students participate in cooperative learning activities, in which communication is essential for success.
- Have students research space communication.

Closure
In their space journals, have students write about how difficult it is sometimes to communicate with others and/or follow directions.

What Would Living and Working in Space Be Like?

Communicate It *(cont.)*

Draw your structure and your partner's structure and then compare.

Trial 1

Similarities and differences: _____

_____

Trial 2

Similarities and differences: _____

_____

What Would Living and Working in Space Be Like?

Record Keeping

Question
What might it be like to keep detailed records of an experience in space?

Setting the Stage
- Review with your students the Scientific Method, paying special attention to steps 1 and 5.
- Discuss with students how important it is for scientists to keep accurate records of their investigations.

Materials Needed for Each Individual
- a watch
- something to observe (e.g., animal, person, weather, etc.)
- data-capture sheet (page 55)

Procedure *(Student Instructions)*
1. Choose something or someone to observe.
2. You will need to set aside 20 minutes every day for five days to complete this activity.
3. Using your data-capture sheet, keep accurate notes of what you are observing. Notice how what you are observing may be constantly changing or always staying the same. Write down or draw everything you see.

Extensions
Set up a center where your students can observe a class pet or an object that is constantly changing. Have them keep accurate notes of its movements.

Closure
In their space journals, have students write their impressions of being a scientist in space doing an investigation.

What Would Living and Working in Space Be Like?

Record Keeping *(cont.)*

Use the space below to keep your observations.

Scientist doing investigation: _____

Object or subject being observed: _____

Dates of observation: _____

Day 1: Start Time _____ End Time _____

Day 2: Start Time _____ End Time _____

Day 3: Start Time _____ End Time _____

Day 4: Start Time _____ End Time _____

Day 5: Start Time _____ End Time _____

What Would Living and Working in Space Be Like?

Space Station Creation

Question
What will the new space station look like?

Setting the Stage
- Ask students if they have ever heard of Space Station Freedom?
- Tell students that NASA is coordinating plans for a permanent space station on which astronauts and scientists can live and work. Countries from all over the world are taking part in its development.

Materials Needed for Each Group
- four toilet paper tubes and one paper towel tube
- several sheets of white paper
- 32 popsicle sticks
- 30 straws
- plastic wrap
- white glue
- transparent tape
- scissors
- diagram of space station (page 57)

Procedure *(Student Instructions)*
1. Cover all of the cardboard tubes with white paper.
2. Attach two of the toilet paper tubes side-by-side to the center of the paper towel tube. (See diagram, page 57.)
3. To make one solar panel, you will need three popsicle sticks, six straws, and white glue. Break one of the popsicle sticks in half and glue sticks into a rectangle. Allow time for the popsicle sticks to dry. Next, cut straws to fit the width of the rectangle and glue them inside the rectangle so the final design resembles a ladder. When your ladder has dried, cover it in plastic wrap and it will resemble a solar panel. Repeat this step until you have made eight solar panels. (See diagram, page 57.)
4. Break four popsicle sticks in half and glue each piece to one end of each solar panel. (See diagram, page 57.)
5. Glue two popsicle sticks together so they form a straight line. Repeat this a second time and then glue each set of sticks to the opposite ends of the space station. (See diagram, page 57.)
6. Cut four slits large enough to fit a popsicle stick in each of the remaining toilet paper tubes. Two slits should be made on each side of the tube at the 1" (2.5 cm) mark and the 3" (7.5 cm) mark.
7. Tape or glue each of the solar panels into one of the slits and allow to dry.
8. Glue the solar panel tubes to the popsicle sticks coming out of the body of the space station. Allow time for drying, and your station is completed.

Extensions
- Have students create additional parts to their space stations.
- Have students make an orbiter and dock it to their space stations.

Closure
In their space journals, have students write about what it might be like to live and work on a space station orbiting the earth.

What Would Living and Working in Space Be Like?

Space Station Creation *(cont.)*

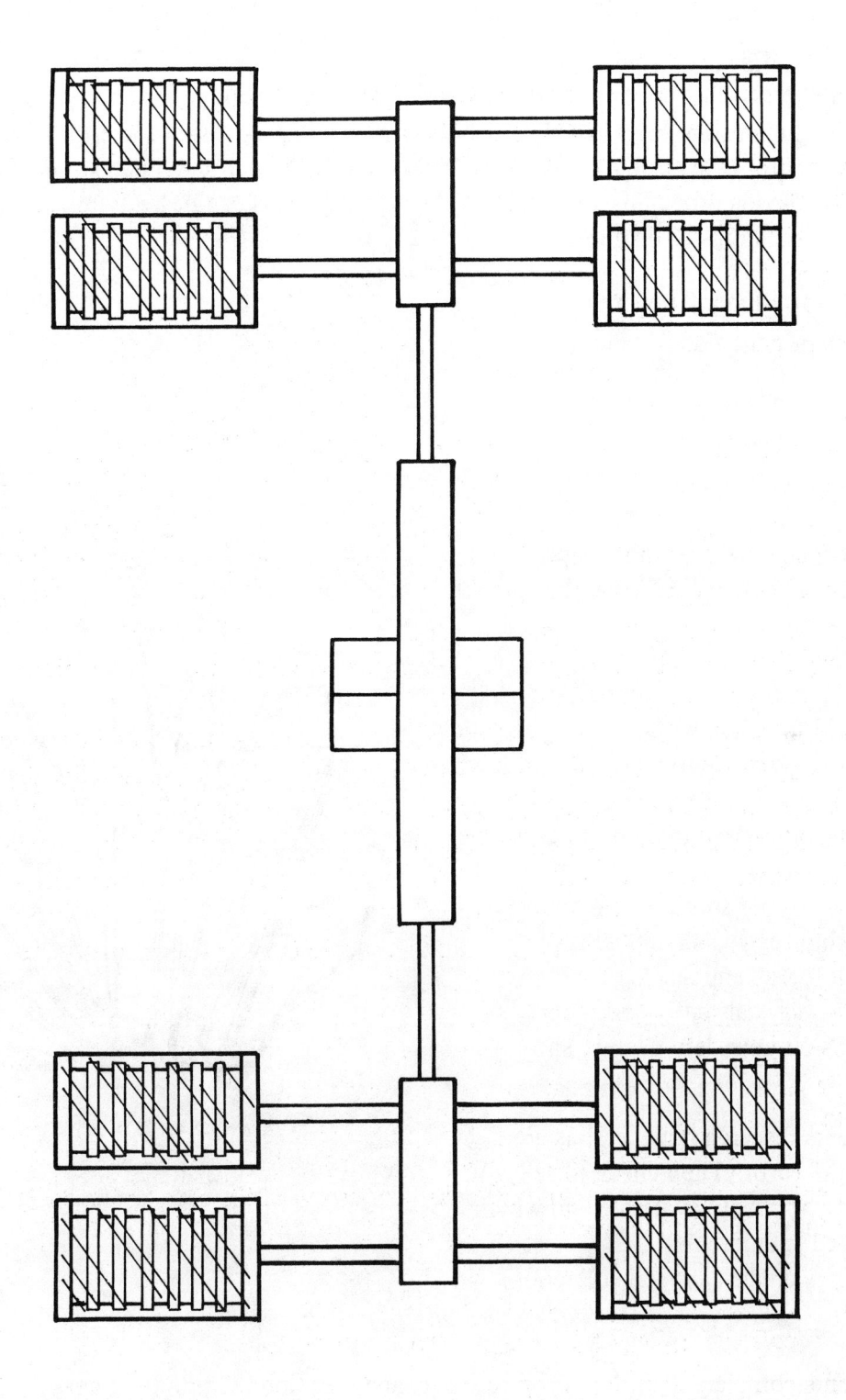

What Would Living and Working in Space Be Like?

Crystal Growth

Question
How do you grow crystals?

Setting the Stage
- Have students define the term *crystal*.
- Have on display a variety of crystals or pictures of crystals for students to examine.
- Discuss with students the importance of crystal growth aboard the space shuttle missions. Crystals naturally come from the earth but take many years to form. After extended studies on board the space shuttle, payload specialists have determined that crystals can be grown in artificial environments much faster than they occur naturally on earth.

Materials Needed for Each Group
- shallow dish or petri dish
- hot water
- Epsom salt
- spoon or stirring stick
- food coloring in dropper bottles
- measuring spoons
- clear disposable 6 oz. (180 mL) cups
- data-capture sheet (page 59), one per student

Procedure (Student Instructions)
1. In a cup dissolve 2 oz. (55 g) of Epsom salt into 2 oz. (60 mL) hot water. When the solid has dissolved, continue to add salt until it will no longer dissolve in the solution.
2. Add food coloring to your solution to give your crystals color.
3. Carefully pour your solution into your shallow dish or petri dish and let it sit undisturbed for several days.
4. Observe your crystals daily and record your observations on your data-capture sheet.

Extensions
- Have students research how crystals are formed in the earth. Then check to see which crystals are commonly available and which ones are hard to find.
- Have students create a crystal collection.

Closure
Have students complete their data-capture sheets and then add them to their space journals.

What Would Living and Working in Space Be Like?

Crystal Growth *(cont.)*

Observe your crystals each day and fill in the information needed.

1. Draw a picture of your crystal and describe what you see after day 1.

2. Draw a picture of your crystal and describe what you see after day 2.

3. Draw a picture of your crystal and describe what you see after day 3.

What Would Living and Working in Space Be Like?

New Worlds

Question

What would it be like to land on another planet with an unfamiliar surface?

Setting the Stage
- Discuss with students that with today's technology we are capable of landing a person on another planet.
- Discuss with students that there are different types of vehicles for different types of terrain.
- Discuss with students necessary adaptations of vehicles so they are able to move on different types of surfaces—for example, chains to drive in the snow.

Materials Needed for Each Group
- bowl or pie tin
- premixed, two parts corn starch to one part water
- an assortment of materials to use as landing testers (e.g., paper towel tubes, pieces of wood, small plastic bottles, toilet paper tubes, etc.)
- colored markers or crayons
- data-capture sheet (page 61), one per student

Procedure *(Student Instructions)*
1. Before you begin testing your landing materials, feel the mixture on which you will be attempting to land.
2. Now that you know the consistency of your surface, begin testing your landing materials to see if they will sink.
3. Record on your data-capture sheet what happens to each landing tester.
4. After each test, try to come up with an adaptation to make your landing tester work.
5. Again record your findings on your data-capture sheet after each follow-up test.

Extensions

Have students research the surface of the remaining eight planets in our solar system. Are any of these surfaces landable? Have them share their information with the class.

Closure

In their space journals, have students write a story about a fictitious planet, where they need to use special adaptations in order to walk on the surface.

What Would Living and Working in Space Be Like?

New Worlds *(cont.)*

Answer the questions below.

1. What are some factors you must know to land safely on this surface?
   _____
   _____

2. How would you go about landing on this type of surface?
   _____
   _____

3. What types of materials were you able to use to land without sinking?
   _____
   _____

4. Write down all your attempts at landing on this surface. What were some of the problems you encountered?
   _____
   _____

5. Draw a picture of what the surface looks like to you.

Curriculum Connections

Language Arts

Reading, writing, listening, and speaking experiences blend easily with the teaching and reinforcement of science concepts. Science can be a focal point as you guide your students through poems and stories, stimulating writing assignments, and dramatic oral presentations. If carefully chosen, language arts material can serve as a springboard to a space lesson, the lesson itself, or an entertaining review.

There is a wealth of good literature to help you connect your curriculum. Some excellent choices are suggested in the Bibliography (pages 95-96).

Science Concept: *Stars and planets have existed for millions of years.*

Every culture in the world has a myth or story about the constellations in the sky. These stories may tell how the stars came to be, what their importance is to a particular culture, or just a story told to entertain. Read to students a story of a constellation (Andromeda, Sagittarius, etc.) and then have them create a constellation of their own and a story to go with it. An example is given below.

Andromeda

Andromeda was the beautiful daughter of the king and queen of Ethiopia. Andromeda's mother bragged that Andromeda was even more beautiful than the goddesses of the sea. When the goddesses of Olympus heard this, they were furious and asked Poseidon, the god of the sea, for revenge. Poseidon sent a great sea monster named Cetus to swim along the shores of the kingdom, devouring everyone he could. Horrified, Andromeda's father pleaded with Poseidon to stop the monster. Poseidon told the king that the only way to save his kingdom was to give Andromeda to Cetus. With no other choice, the king sadly chained his daughter to a rock by the sea and left her there. Luckily for Andromeda, a hero named Perseus happened to be riding by on his flying horse called Pegasus. He bravely killed the sea monster and carried the beautiful princess away.

Curriculum Connections

Language Arts

Science Concept: *Planets have different surfaces.*

In a large bowl, have students mix two parts corn starch with one part water. Tell them that this mixture will feel and look like the surface they will be landing on when they reach their new planet. Have them write and illustrate how they will survive and travel on this new surface.

©1994 Teacher Created Materials, Inc. #634 Space

Curriculum Connections

Social Studies

As you guide your students through lessons in history, geography, cultural awareness, or other areas of social studies, keep in mind the role space studies can play. You will find it easy to incorporate the teaching and reinforcement of science concepts in your lessons.

Science Concept: *Space missions are much more common today.*

- Pretend your classroom is the space shuttle and have students study each continent as they fly over it. Have them make a small paper shuttle and place it on a classroom map to denote what country they are studying.

- Have students get into groups and make a U.S. map with all of the NASA bases starred.

- Have students write letters to former astronauts, asking questions about space travel, walking on the moon, etc.

- Have students research how spin-off programs of space innovations have changed people's lives in the world, U.S., or just their community.

- Have a guest speaker come to class to talk about careers in aviation.

- Have students write to their congressional leaders supporting or opposing the space program.

Curriculum Connections

Physical Education

What can be more fun for primary students then pretending they are astronauts flying through space. Here is an opportunity to let your students develop their knowledge of space in a physical way.

Science Concept: *Parachutes were used in the landing process of early space capsules.*

- Have students play in a gym or outside with a large parachute. One game that can be played is called the parachute toss. In this game the entire class must circle around a "parachute" and hold it with both hands. After placing a ball in the center of the parachute, they must toss it up in the air as many times as possible without touching the ball or letting it hit the ground. The number of consecutive successful bounces is the score for this activity. Teamwork and communication are much more important than the actual score.

- The game "Duck, Duck, Goose," can be easily modified into "Astronaut, Astronaut, Alien."

©1994 Teacher Created Materials, Inc. #634 Space

Curriculum Connections

Math

The study of space requires the use of math-oriented skills. Measuring, comparing, and graphing are just a few of the skills that can bring mathematics into your space lessons.

- Teach or review the use of measuring tools (such as rulers with centimeters and inches to measure length).

- Have students practice reading and making charts and graphs.

- Provide opportunities for students to record data on a variety of graphs and charts. Teach the skills necessary for success.

- Encourage students to devise their own ways to show the data they have gathered.

- On an appropriate level, teach how to average test results.

- Challenge students to find mathematical connections as they study space.

Science Concept: *There are nine planets in our solar system.*

Have students calculate the distances and the circumferences of several planets in our solar system. This may require additional explanation.

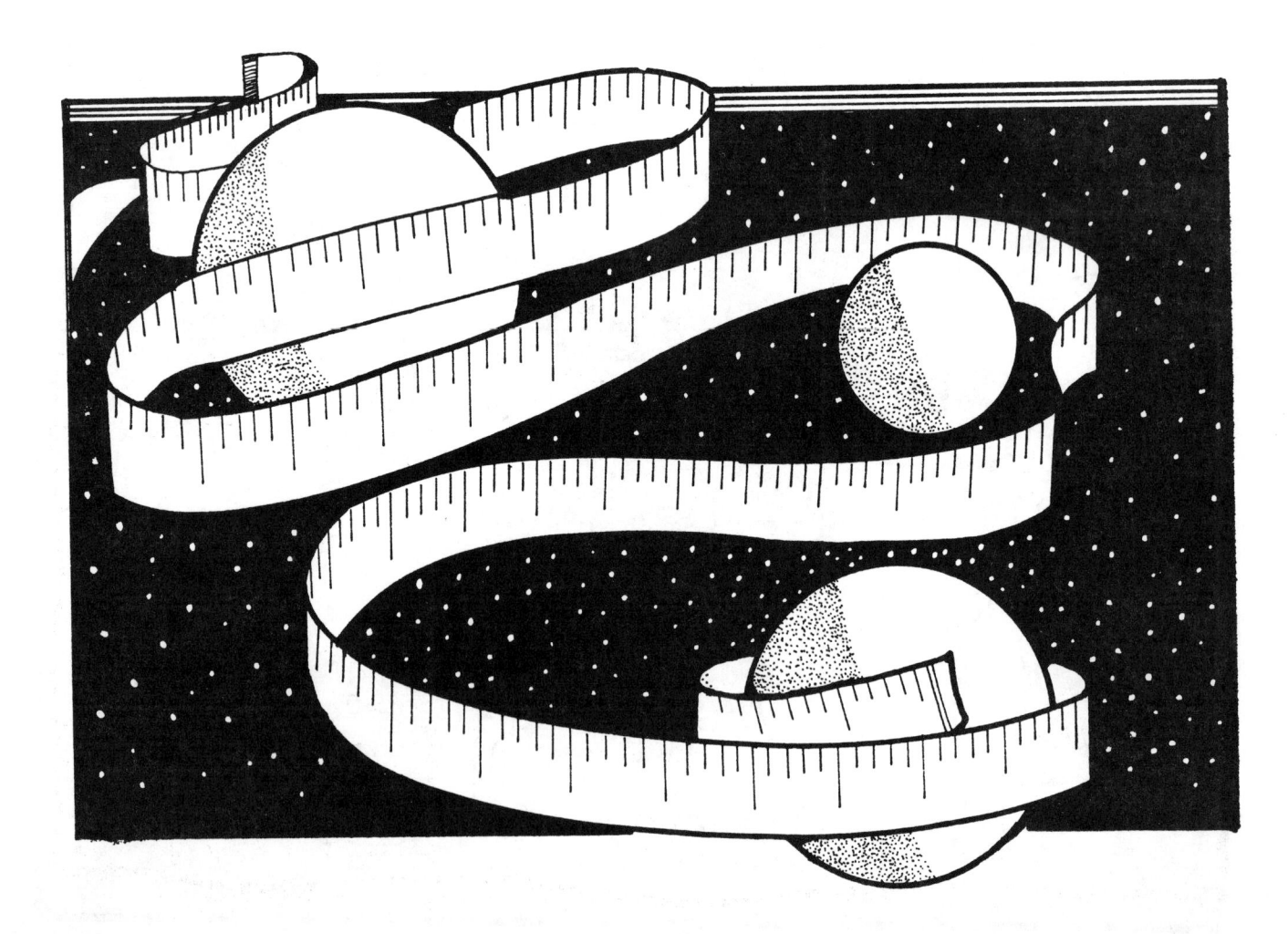

Curriculum Connections

Art

Art projects using space as the theme are excellent ways to extend any space lesson. Using items from home or school will give your students endless "space possibilities!"

Science Concept: *Space stations will become common in the future.*

- Have students create a space city of their own with materials they bring in from home or things they can find at school.

- Have the entire class create a mural representing the planets of our solar system, the sun, satellites, and a space shuttle flying a mission.

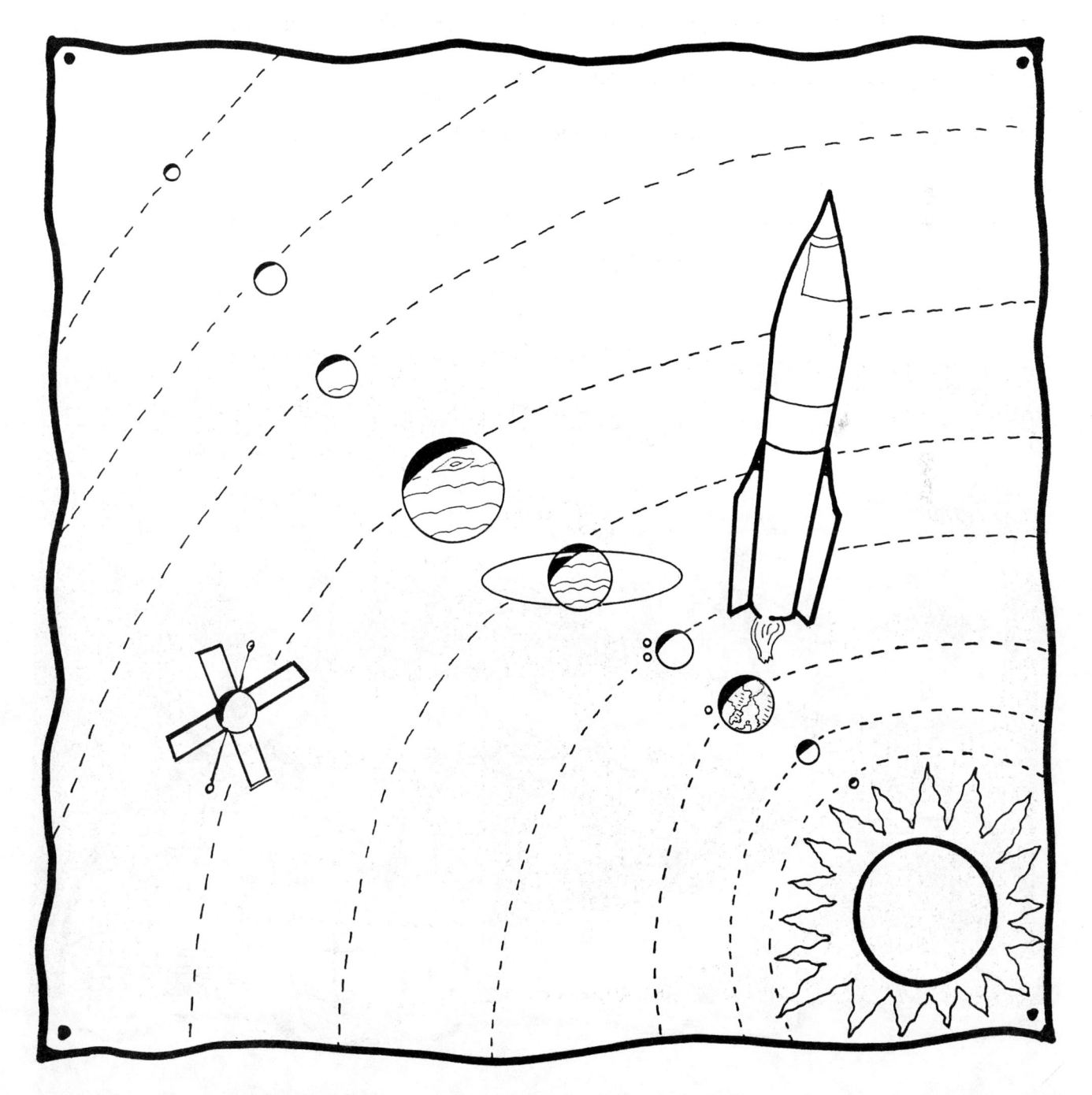

©1994 Teacher Created Materials, Inc.   67   #634 Space

Curriculum Connections

Music

Singing songs about space, selecting orchestral numbers to promote space education, and making instruments are just a few of the ways to integrate music into your space-based lessons.

Science Concept: *The theme of space has been used in music and theater.*

- Have students pick one of their favorite songs from the radio, television, or theater and change the lyrics to something space-related.

- Have students create a musical or play about actual space exploration or fantasy exploration.

Teacher Note: See page 10 for instructions. *Station-to-Station Activity*

Mission Alpha

Before beginning your investigation, write your group members' names by their jobs on the lines below.

_____ Mission Leader   _____ Specialist 2

_____ Specialist 1   _____ Specialist 3

Fact: To release a satellite from the cargo bay, astronauts use a special remote manipulator arm.

Materials Needed for This Mission
- tongs
- building blocks
- color markers or crayons
- data-capture sheet (page 70), one per student

Procedure *(Student Instructions)*

1. Your mission, should you accept, is to build a structure with the instruments provided.
2. Using the tongs, you are to build a structure out of the building blocks. You may not touch anything with your hands. You must use the tongs for everything you do.
3. When your structure is complete, draw a picture in the space provided on your data-capture sheet.
4. Make sure you name your structure.
5. Clean up your area before you move to the next station.
6. Put your finished data-capture sheet in the collection pocket on the side of the table at this station.

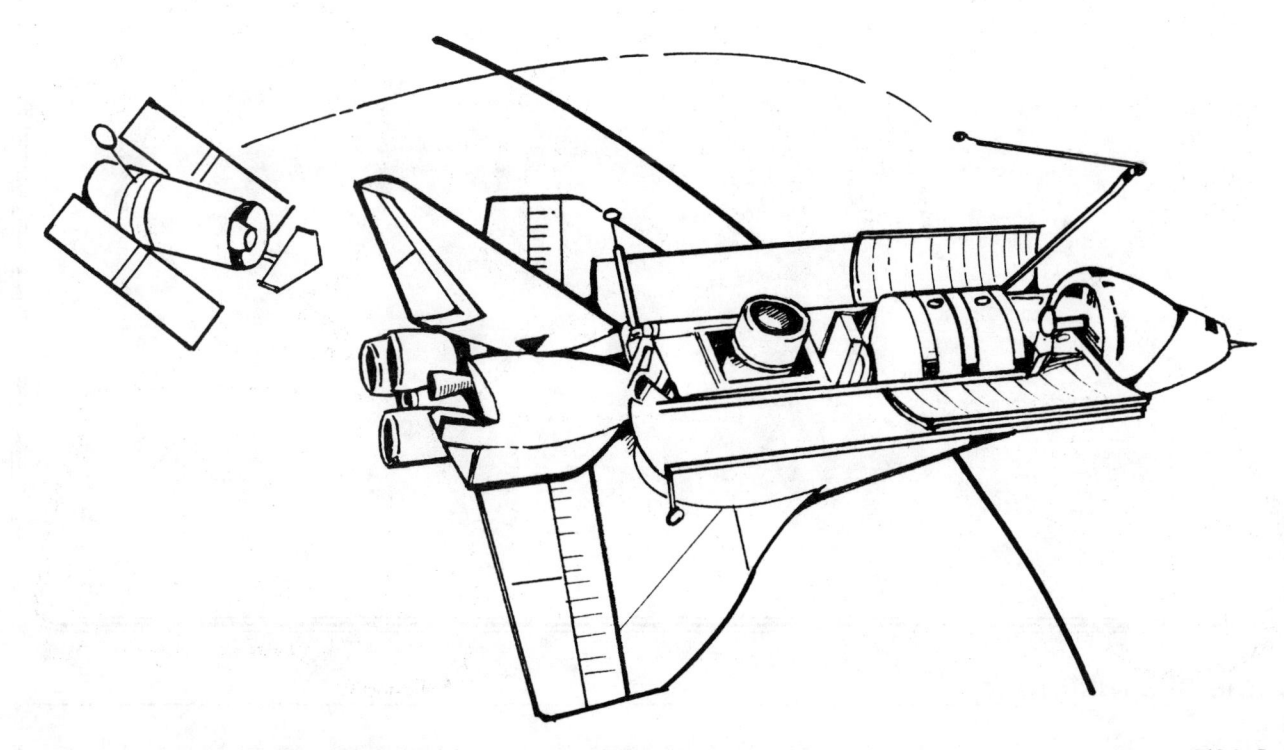

©1994 Teacher Created Materials, Inc.   #634 Space

Station-to-Station Activity

Mission Alpha *(cont.)*

Draw and name your structure below.

Name of structure: _____

Teacher Note: See page 10 for instructions.   *Station-to-Station Activity*

Mission Beta

Before beginning your investigation, write your group members' names by their jobs on the lines below.

_____ Mission Leader   _____ Specialist 2

_____ Specialist 1   _____ Specialist 3

Fact: The gloves that are part of the new astronaut suits are the only customized portion of the suit. Each pair of gloves will fit only one astronaut.

Materials Needed for This Mission
- several pairs of gloves (e.g., gardening, mittens, rubber, snow, etc.)
- medium size container—for example, a cut-off milk carton containing: brads, paper clips, pennies, pieces of cereal, rubber bands, small pencil, and stick pin
- brown bottle containing: assorted math manipulatives, bobby pins/hair pins, and small pieces of paper
- small cardboard box, wrapped—for example, a shoe box containing anything that will fit in it
- medium to large cardboard box
- data-capture sheet (page 72), one per student

Procedure *(Student Instructions)*
1. Choose a set of gloves from what is available and put them on.
2. While wearing a pair of gloves, pick up and empty the white container into a box. Sort the objects you see into different piles. Then, describe and draw how you accomplished this task.
3. Next, empty the brown bottle and then put the items back into the bottle. Record which item was the easiest, which item was the hardest, and why.
4. Finally, unwrap the small box, look at the object inside, wrap the box back up, and draw a picture of the object you observed.
5. Put your finished data-capture sheet in the collection pocket on the side of the table at this station.

©1994 Teacher Created Materials, Inc.   #634 Space

Station-to-Station Activity

Mission Beta (cont.)

Fill in the information needed.

1. Describe and draw how you emptied and sorted the items in the container.

2. Which was the easiest object to put back into the brown bottle? Explain.

_____

_____

Which was the hardest object to put back into the brown bottle? Explain.

_____

_____

3. Draw a picture of what you saw inside the covered box.

Teacher Note: See page 10 for instructions.   *Station-to-Station Activity*

Mission Gamma

Before beginning your investigation, write your group members' names by their jobs on the lines below.

_____ Mission Leader   _____ Specialist 2

_____ Specialist 1   _____ Specialist 3

Fact: The orbiter lands back on earth by gliding to the landing strip.

Materials Needed for This Mission
- a series of pictures showing the different phases of the space shuttle (e.g., launch pad, lift-off, orbiting in space, releasing cargo, re-entry, and landing)
- colored markers or crayons
- data-capture sheets (pages 74-75), one each per student

Procedure *(Student Instructions)*
1. You will be drawing and labeling several stages of a space shuttle mission.
2. Draw and label the space shuttle on the launch pad.
3. Draw and label the space shuttle at lift-off or separation.
4. Draw and label the orbiter orbiting in space.
5. Draw and label the orbiter unloading cargo from its cargo bay.
6. Draw and label the orbiter on re-entry into the earth's atmosphere.
7. Draw and label the orbiter landing back on earth.
8. Put your finished data-capture sheets in the collection pocket on the side of the table at this station.

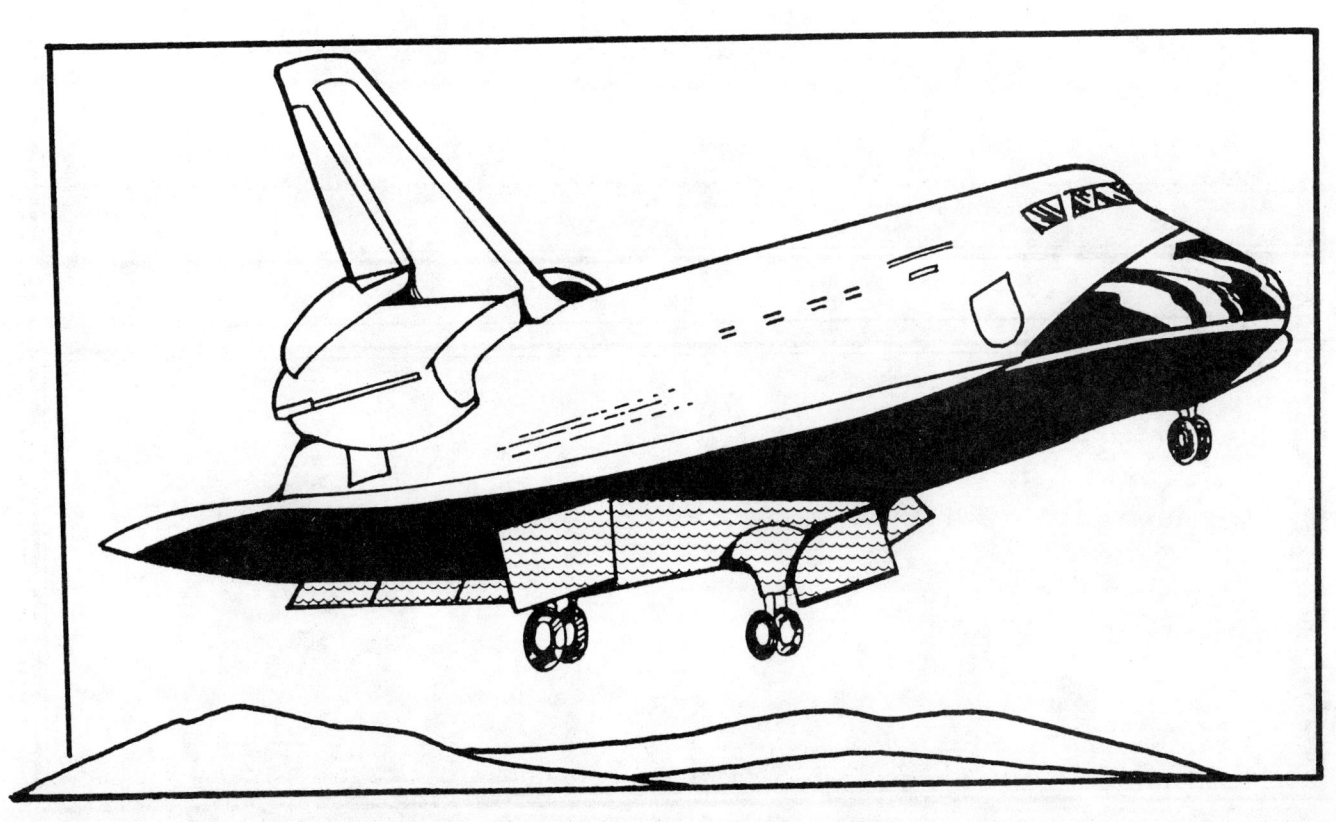

©1994 Teacher Created Materials, Inc.   #634 Space

Station-to-Station Activity

Mission Gamma (cont.)

Draw and label the stages of the space shuttle.

#1: _____

#2: _____

#3: _____

Station-to-Station Activity

Mission Gamma *(cont.)*

Draw and label the stages of the space shuttle.

#4: _____

#5: _____

#6: _____

©1994 Teacher Created Materials, Inc. 75 #634 Space

Station-to-Station Activity		Teacher Note: See page 10 for instructions.

Mission Mu

Before beginning your investigation, write your group members' names by their jobs on the lines below.

_____ Mission Leader   _____ Specialist 2

_____ Specialist 1   _____ Specialist 3

Fact: NASA is the National Aeronautics and Space Administration.

Materials Needed for This Mission
- colored markers or crayons
- data-capture sheet (page 77), one per student

Procedure *(Student Instructions)*
1. On top of your stationery, draw a picture of something space-related.
2. Next, you are to write a thank-you letter to NASA. You might want to thank them for any information and materials they may have sent to your class.
3. Put your finished data-capture sheet in the collection pocket on the side of the table at this station.

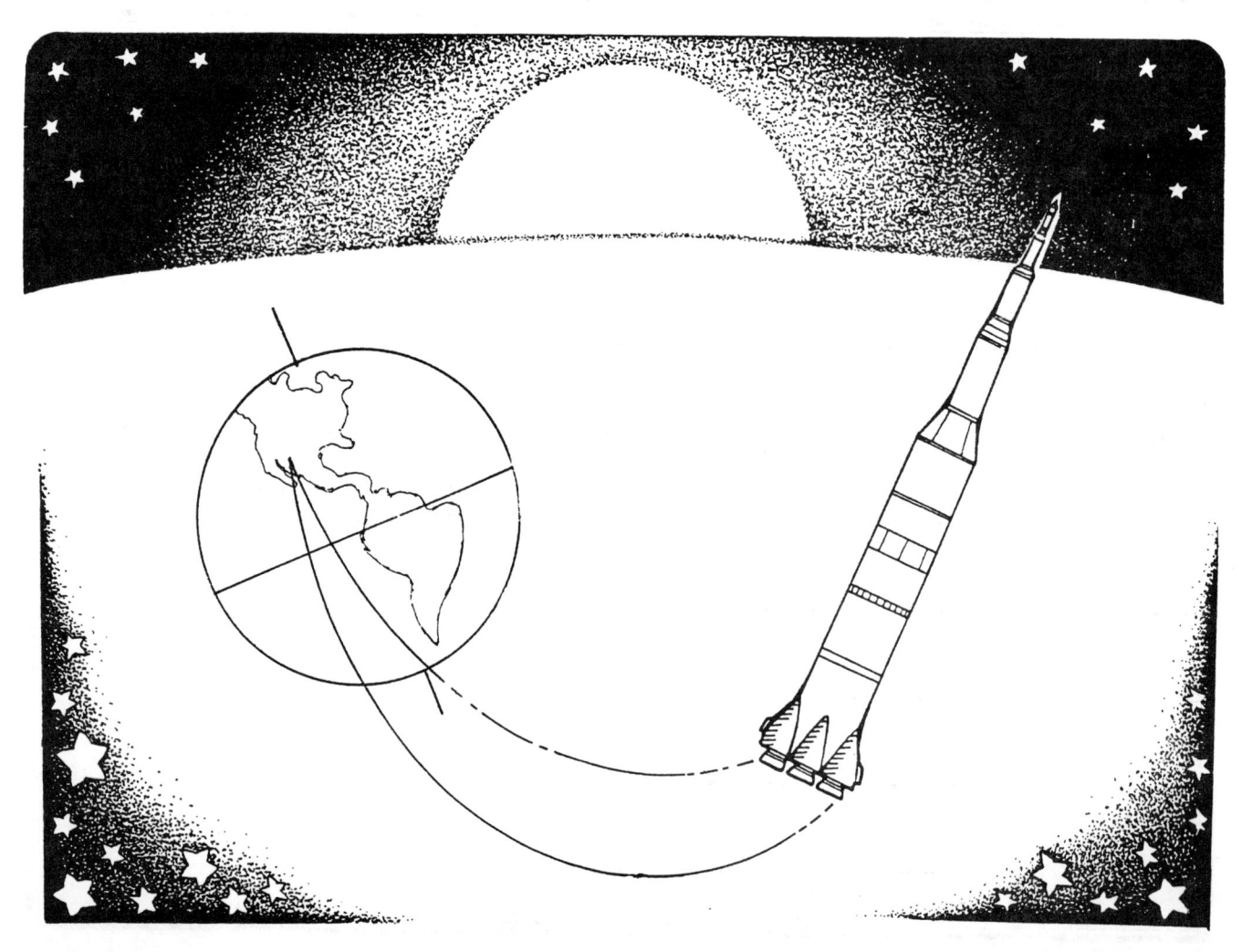

Station-to-Station Activity

Mission Mu *(cont.)*

Station-to-Station Activity   Teacher Note: See page 10 for instructions.

Mission Phi

Before beginning your investigation, write your group members' names by their jobs on the lines below.

_____ Mission Leader   _____ Specialist 2

_____ Specialist 1   _____ Specialist 3

Fact: A telescope makes objects at a distance appear to be closer than they really are. The Hubble Space Telescope is orbiting the earth.

Materials Needed for This Mission
- a variety of specimens (e.g., cinnamon, dirt, feathers, gravel, hair, sand, salt, sugar)
- magnifying lens
- macroscope or microscope
- data-capture sheets (pages 79–80), one each per student

Procedure *(Student Instructions)*
1. Choose six of the available specimens.
2. You will be drawing three pictures of each specimen.
3. Draw your first picture while looking at your specimen with normal eyesight.
4. Draw your second picture while looking at your specimen with a magnifying lens.
5. Draw your third picture while looking at your specimen with a macroscope or microscope.
6. Repeat steps 3–5 for each new specimen.
7. Put your finished data-capture sheets in the collection pocket on the side of the table at this station.

#634 Space   ©1994 Teacher Created Materials, Inc.

Station-to-Station Activity

Mission Phi *(cont.)*

Draw each specimen three times.

Normal Eye View	Magnifying Lens	Macro/Microscope View

Station-to-Station Activity

Mission Phi *(cont.)*

Draw each specimen three times.

Normal Eye View	Magnifying Lens	Macro/Microscope View

#634 Space

Management Tools

Science Safety

Discuss the necessity for science safety rules. Reinforce the rules on this page or adapt them to meet the needs of your classroom. You may wish to reproduce the rules for each student, or post them in the classroom.

1. Begin science activities only after all directions have been given.

2. Never put anything in your mouth unless it is required by the science experience.

3. Always wear safety goggles when participating in any lab experience.

4. Dispose of waste and recyclables in proper containers.

5. Follow classroom rules of behavior while participating in science experiences.

6. Review your basic class safety rules every time you conduct a science experience.

You can still have fun and be safe at the same time!

Management Tools

Space Journal

Space Journals are an effective way to integrate science and language arts. Students are to record their observations, thoughts, and questions about past science experiences in a journal to be kept in the science area. The observations may be recorded in sentences or sketches which keep track of changes both in the science item or in the thoughts and discussions of the students.

Space Journal entries can be completed as a team effort or an individual activity. Be sure to model the making and recording of observations several times when introducing the journals to the science area.

Use the student recordings in the Space Journals as a focus for class science discussions. You should lead these discussions and guide students with probing questions, but it is usually not necessary for you to give any explanation. Students come to accurate conclusions as a result of classmates' comments and your questioning. Space Journals can also become part of the students' portfolios and overall assessment program. Journals are valuable assessment tools for parent and student conferences as well.

How To Make a Space Journal

1. Cut two pieces of 8 ½" x 11" (22 cm x 28 cm) construction paper to create a cover. Reproduce page 83 and glue it to the front cover of the journal. Allow students to draw space pictures in the box on the cover.
2. Insert several Space Journal pages. (See page 84.)
3. Staple together and cover stapled edge with book tape.

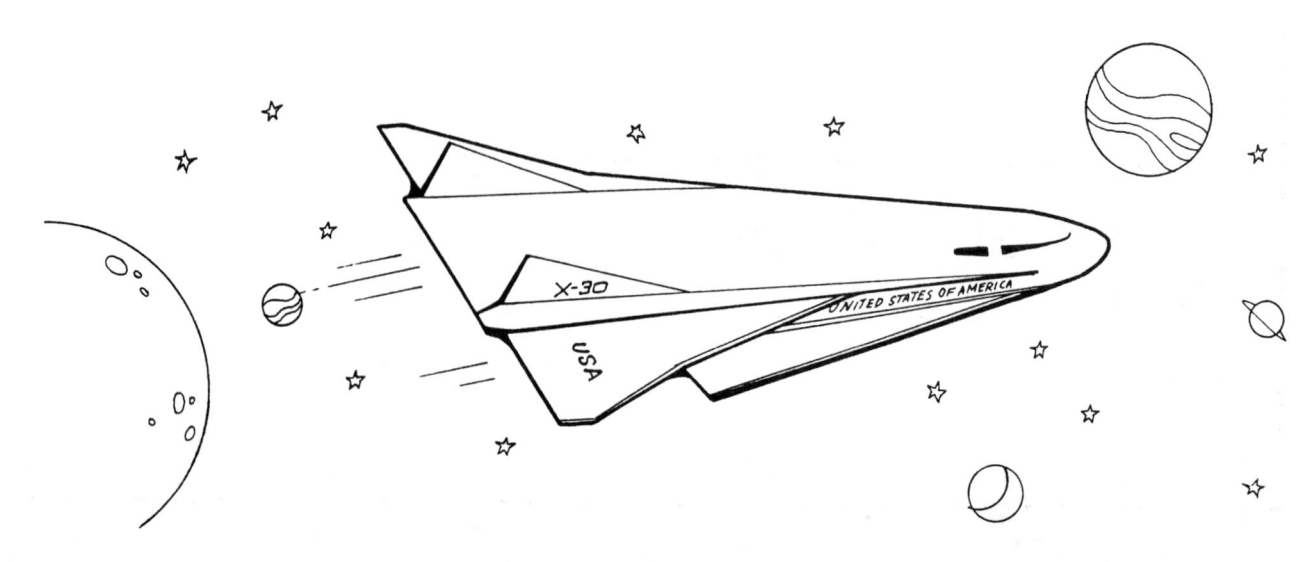

Management Tools

My Space Journal

Name _____

Management Tools

Space Journal

Illustration

This is what happened: _____
_____
_____
_____
_____

This is what I learned: _____
_____
_____
_____
_____

Management Tools

My Science Activity

K-W-L Strategy

Answer each question about the topic you have chosen.

Topic: _____

K - What I Already **Know:** _____

_____

_____

_____

_____

W - What I **Want to Find Out:** _____

_____

_____

_____

_____

L - What I **Learned After Doing the Activity:** _____

_____

_____

_____

_____

Management Tools

Investigation Planner *(Option 1)*

Observation

_____

_____

Question

_____

_____

Hypothesis

_____

_____

Procedure

Materials Needed:

_____

_____

_____

Step-by-Step Directions: (Number each step!)

_____

_____

_____

_____

_____

_____

Management Tools

Investigation Planner *(Option 2)*

Science Experience Form

Scientist _____

Title of Activity _____

Observation: What caused us to ask the question?
_____
_____
_____

Question: What do we want to find out?
_____
_____
_____

Hypothesis: What do we think we will find out?
_____
_____
_____

Procedure: How will we find out? *(List step by step.)*

1. _____
2. _____
3. _____
4. _____

Results: What actually happened?
_____
_____
_____

Conclusions: What did we learn?
_____
_____

Management Tools
Space Observation Area

In addition to station-to-station activities, students should be given other opportunities for real-life science experiences. For example, model rockets can provide a vehicle for discovery learning if students are given time and space to observe them.

Set up a space observation area in your classroom. As children visit this area during open work time, expect to hear stimulating conversations and questions among them. Encourage their curiosity, but respect their independence!

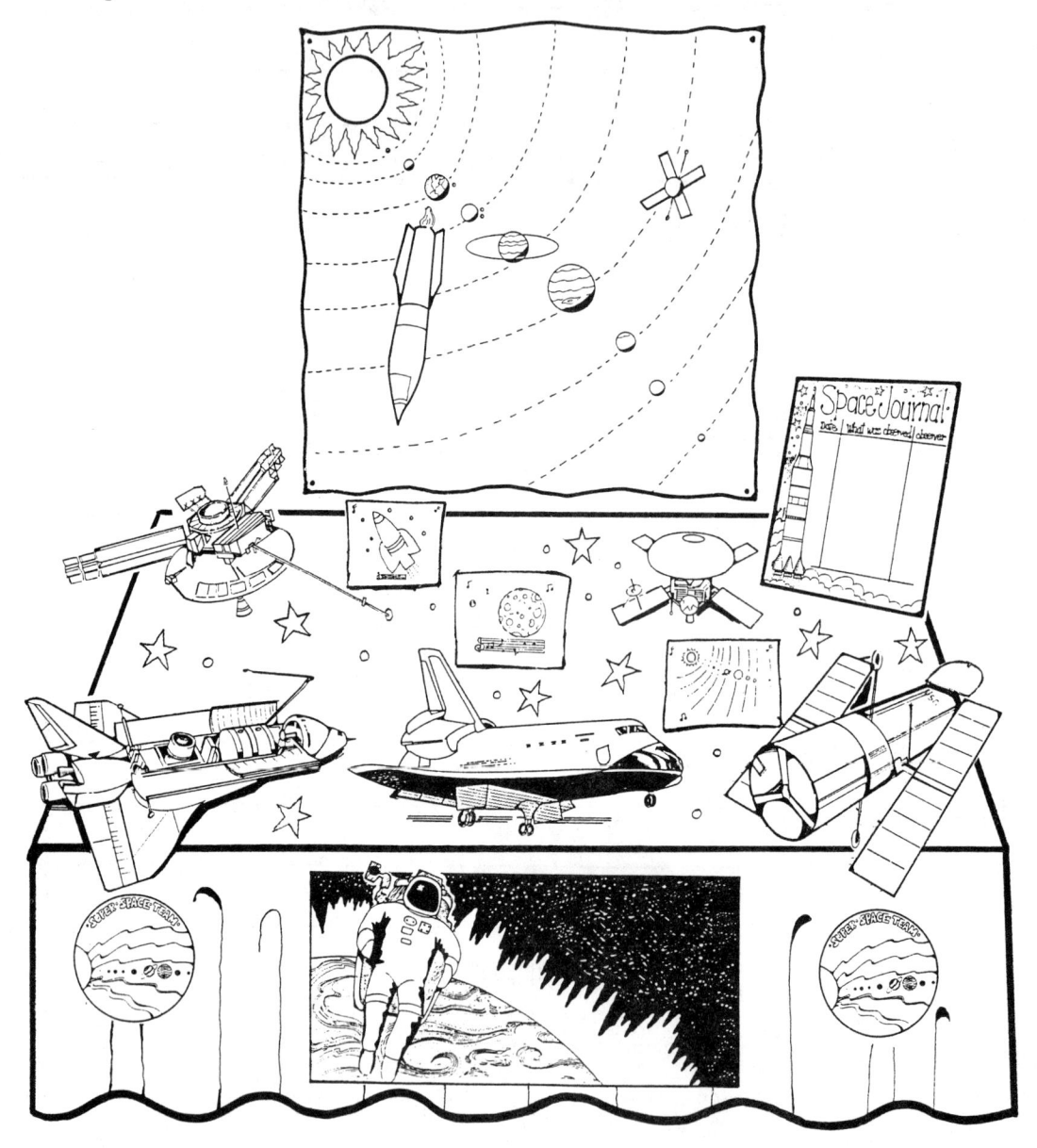

Books with facts pertinent to the subject, item, or process being observed should be provided for students who are ready to research more sophisticated information.

Sometimes it is very stimulating to set up a science experience or add something interesting to the Space Observation Area without a comment from you at all! If the experiment or materials in the observation area should not be disturbed, reinforce with students the need to observe without touching or picking up.

#634 Space ©1994 Teacher Created Materials, Inc.

Management Tools

Assessment Forms

The following chart can be used by the teacher to rate cooperative-learning groups in a variety of settings.

Science Groups Evaluation Sheet

Room: _____   Date: _____

Activity: _____

Everyone	**Group**									
	1	2	3	4	5	6	7	8	9	10
. . . gets started.										
. . . participates.										
. . . knows jobs.										
. . . solves group problems.										
. . . cooperates.										
. . . keeps noise down.										
. . . encourages others.										

Teacher comment

_____

Bragging rights for the group session: _____

_____

©1994 Teacher Created Materials, Inc.   #634 Space

Management Tools

Assessment Forms *(cont.)*

The evaluation form below provides student groups with the opportunity to evaluate the group's overall success.

Cooperative Group Evaluation

Assignment: _____

Date: _____

Scientists	Jobs
_____	_____
_____	_____
_____	_____
_____	_____

As a group, decide which face you should fill in and complete the remaining sentences.

1. We finished our assignment on time, and we did a good job.

2. We encouraged each other, and we cooperated with each other.

3. We did best at _____
   _____.

4. Next time we could improve at _____
   _____.

#634 Space 90 ©1994 Teacher Created Materials, Inc.

Management Tools

Assessment Forms *(cont.)*

The following form may be used as part of the assessment process for hands-on science experiences.

Science Anecdotal Record Form

Date: _____

Scientist's Name: _____

Topic: _____

Assessment Situation: _____

Instructional Task: _____

Behavior/Skill Observed: _____

_____

_____

_____

_____

_____

This behavior/skill is important because _____

_____

_____

_____

_____ .

Management Tools

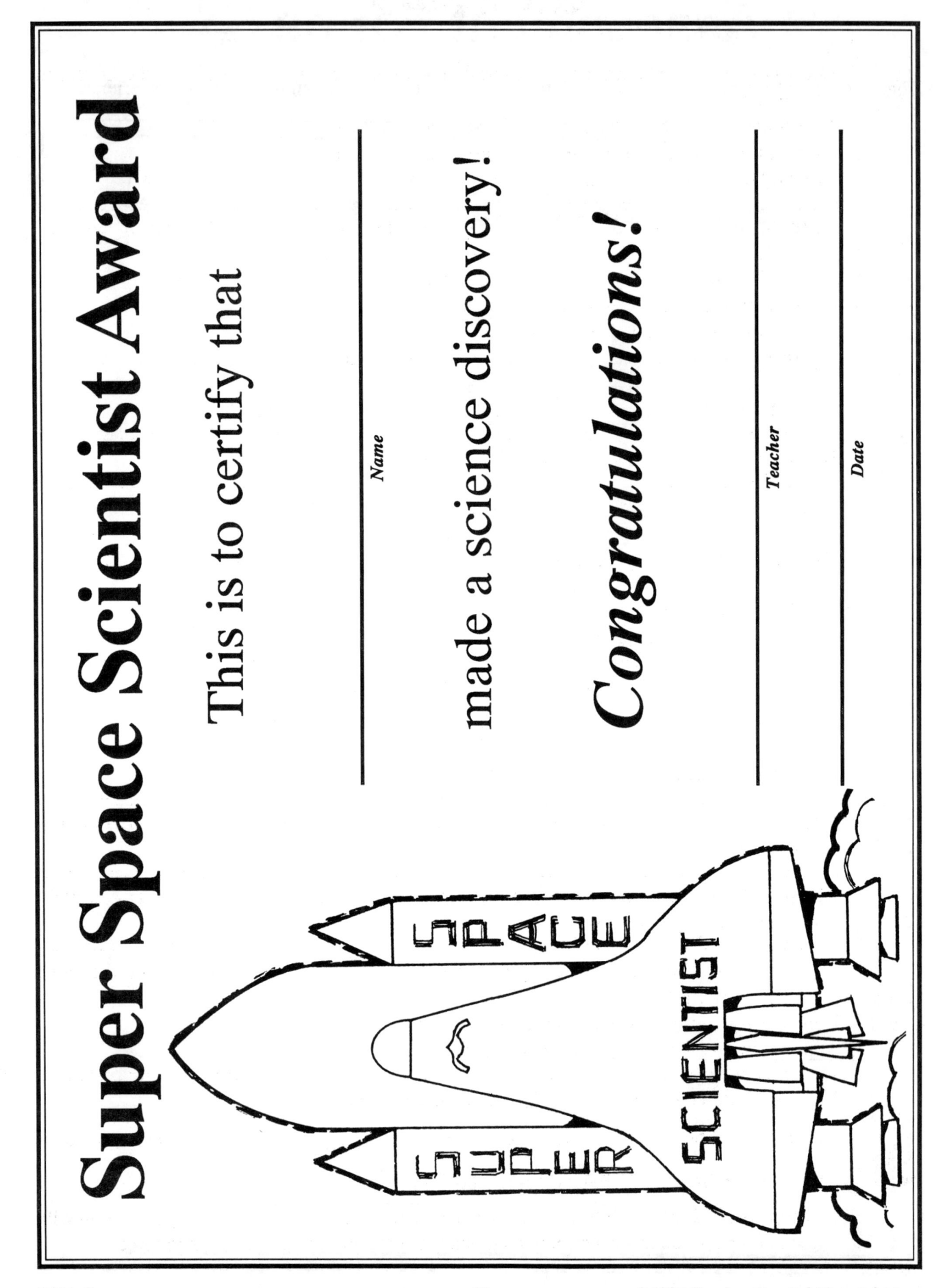

Super Space Scientist Award

This is to certify that

_____
Name

made a science discovery!

Congratulations!

_____
Teacher

_____
Date

#634 Space 92 ©1994 Teacher Created Materials, Inc.

Glossary

A

Aerodynamics—the branch of physics dealing with the force exerted by air or other gasses in motion.
Aerospace—everything that is above the surface of the earth.
Astronomy—the science of the stars, planets, their motion, size, etc.
Atmosphere—layer of gasses above the earth or any other planet.

C

Centrifugal Force—the force in a circle that wants the object to escape the circle in which it is traveling.
Circumference—the curve or boundary of a circle.
Conclusion—the outcome of an investigation.
Constellation—a group of fixed stars.
Control—a standard measure of comparison in an experiment. The control always stays constant.
Crystal—a solid that is made up of atoms which are set in a systematic pattern.

D

Diameter—the inside measurement of a circle from its two widest points.

E

Eclipse—the darkening of a celestial body.
Elliptical—having the form of the ellipse, the shape of an oval.
Escape Velocity—the speed necessary to break away from a celestial body. For earth this is about 25,000 miles/hour (40,000 km/hour).
EVMU—extra vehicular mobility unit (space suit).
Experiment—a means of proving or disproving a hypothesis.

G

Gravity—the pull on all bodies towards the earth's center.

Friction—resistance to motion of surfaces that touch.

Galaxy—any large system of stars, like the Milky Way.

H

Hypothesis (hi-POTH-e-sis)—an educated guess to a question for which one is trying to find the answer.

I

Investigation—observation of something followed by a systematic inquiry in order to explain what was originally observed.

M

Meteor—a small solid body entering the earth's atmosphere from outer space at a great speed (shooting star).
Meteorite—a stone or metal mass remaining from a meteor fallen to earth.
Momentum—the mass of an object times its velocity.

N

NASA—National Aeronautics and Space Administration.
Natural Satellite—a smaller body, such as a planet, moving around a larger body or planet.
Neil Armstrong—first man to set foot on the moon (July 20, 1969).
Newton's First Law of Motion—a body at rest will remain at rest, and a body in motion will remain in motion at a constant velocity unless acted upon by an unbalanced force.

©1994 Teacher Created Materials, Inc.

Glossary (cont.)

N
Newton's Second Law of Motion—force equals mass times acceleration.
Newton's Third Law of Motion—for every action there is an opposite and equal reaction.

O
Observation—careful notice or examination of something.
Orbit—the path of an object around another body.
Orbiting Speed—the rate that an object moves in a curved path.

P
Period of Rotation—time required to spin around an axis that passes through the body.
Period of Revolution—the time it takes one body to move around another body.
Planet—any heavenly body that revolves around the sun.
Procedure—the series of steps carried out when doing an experiment.

Q
Question—a formal way of inquiring about a particular topic.

R
Radius—the measurement from the middle of the circle to the circumference.
Reflection—the bouncing of a wave from the surface.
Refraction—the change in the speed of light as it moves out of one body and into another.
Results—the data collected after performing an experiment.
Revolve—to move around a central point.
Rotate—to spin on an axis.

S
Satellite—a small object that circles a larger body.
Scientific Method—a creative and systematic process of proving or disproving a given question, following an observation. Observation, question, hypothesis, procedure, results, conclusion, and future investigations comprise the scientific method.
Scientific-Process Skills—the skills needed to be able to think critically. Process skills include: observing, communicating, comparing, ordering, categorizing, relating, inferring, and applying.
Space—the area through which all objects in the universe travel.
Space Transportation System—the orbiter, main engines, solid rocket boosters, and external fuel tank of the U.S. space transport.
Splashdown—the landing of the early space capsules in the ocean.
Solar Energy—energy from the sun.
Star—a giant sphere of incandescent gas in the sky.
Sun—a incandescent body of gases about which the earth and the other planets in our galaxy revolve.

V
Variable—the changing factor of an experiment.

Bibliography

Abernathy, Susan. *Space Machines.* Western Pub., 1991.

Asimov, Isaac. *What Makes the Sun Shine?* Little Brown, 1971.

Attmore, Stephen. *Now You Can Read About Space.* Brimax, 1985.

Baird, Anne. *Space Camp: The Great Adventures for NASA Hopefuls.* Morrow Junior Books, 1992.

Barrett, Norman. *Space Shuttle.* Watts, 1985.

Barton, Byron. *I Want to Be an Astronaut.* Crowell, 1988.

Berenstain, Stan, and Jan Berenstain. *The Berenstain Bears' Science Fair.* Random House, 1977.

Blocksma, Mary & Dewey Blocksma. *Space-Crafting: Invent Your Own Flying Spaceships.* Prentice Hall, 1986.

Bonnet, Robert & Daniel Keen. *Space & Astronomy: Forty-Nine Science Fair Projects.* Tab Books, 1991.

Branley, Franklyn M. *Comets.* Harper LB, 1984.

Branley, Franklyn M. *Is There Life In Outer Space?* Harper LB, 1984.

Branley, Franklyn M. *Journey Into the Black Hole.* Harper LB, 1986.

Branley, Franklyn M. *Rockets and Satellites.* Harper LB, 1987.

Branley, Franklyn M. *Space Telescope.* Harper LB, 1985.

Branley, Franklyn M. *Sunshine Makes the Seasons.* Harper LB., 1988.

Branley, Franklyn M. *The Planets In Our Solar System.* Crowell, 1987.

Ciupik, Larry. *The Universe.* Raintree LB, 1978.

Coords, Arthur E. *Space Apple Story: The Children's Tribute to the Seven Challenger Astronauts.* A E Coords, 1992.

Couper & Henbest. *Telescopes and Observatories.* Watts, 1987.

Darling, David J. *The Planets: The Next Frontier.* Dillon, 1984.

Darling, David J. *Other Worlds: Is There Life Out There?* Dillon, 1985.

Fradin, Dennis B. *Astronomy.* Childrens LB, 1983.

Fradin, Dennis B. *Skylab.* Childrens LB, 1984.

Fradin, Dennis B. *Space Colonies.* Childrens, 1985.

Fradin, Dennis. *Spacelab.* Childrens, 1984.

Freeman, Don. *Space Witch.* Puffin Bks, 1979.

Freeman, Mae, & Ira Freeman. *The Sun, the Moon and the Stars.* Random LB, 1979.

Frisky, Margaret. *Space Shuttles.* Childrens LB, 1982.

Gaffney, Tim. *Kennedy Space Center.* Childrens Press, Inc., 1985.

Gibbons, Gail. *Sun Up, Sun Down.* Harcourt, 1983.

Greene, Carol. *Astronauts.* Childrens LB, 1984.

Hamer, Martyn. *The Night Sky.* Watts, 1983.

Hamilton, Sue. *Space Shuttle Challenger's Explosion.* Abdo & Dghtrs, 1989.

Hansen, Rosanna, and Robert Bell. *My First Book About Space.* Simon & Schuster, 1985.

Hawkes, Nigel. *Space Shuttle.* Watts LB, 1983.

Jaspersohn, William. *How the Universe Began.* Watts, 1985.

Jay, Michael. *Space Shuttle.* Watts LB, 1984.

Jefferis, David. *Satellites.* Watts, 1987.

Lambert, David. *The Solar System.* Watts, 1984.

Lewellen, Hohn. *Moon, Sun and Stars.* Childrens LB, 1981.

Bibliography *(cont.)*

Livingston, Myra Cohn. *Space Songs.* Holiday, 1988.
Marshall, Edward. *Space Case.* Dial Bks Young, 1982.
Marzollom, Jean, and Claudio Marzollo. *Jed's Junior Space Patrol.* Dial, 1982.
McKay, David. *Space Scientist's Projects for Young Scientists.* Watts, 1989.
Millspaugh, Ben. *Aviation and Space Science Projects.* McGraw-Hill, 1992.
Moche, Dinah L. *The Astronauts.* Random Paper, 1979.
Petty, Kate. *Satellites.* Franklin Watts, Inc., 1984.
Pienkowski, Jan. *Robot.* Delacorte, 1981.
Pendoendorf, Illa. *Space.* Childrens, 1982.
Rey, H.A. *Curious George Gets a Medal.* Houghton Mifflin, 1957.
Richards, Gregory. *Satellites.* Childrens, 1983.
Ride, Sally & Susan Okie. *To Space and Back.* Lothrop, 1986.
Ridpath, Ian. *Space.* Kingfisher Books, 1992.
Roop, Peter. *Space Out: Jokes About Outer Space.* Lerner Pub, 1984.
Seiger, Barbara. *Seeing Stars*. Calsbeek, 1993.
Simon, Seymour. *Galaxies.* Morrow, 1991.
Vaden, Judy. *Thematic Unit: Flight.* Teacher Created Materials, 1991.
VanCleave's, Janice. *Astronomy for Every Kid.* Wiley and Sons, 1991.
Wheat, Janis. *Let's Go to the Moon.* National Geographic, 1977.

Spanish Titles

Brown, M. *Buenas noches luna (Good Night Moon).* Lectorum, 1990.
DeBrunhoff, L. *Barbar visita otro planeta (Babar Visits Outer Space).* Lectorum, 1990.
Ehlert, L. *Un lazo a la luna (Moon Rope).* Harcourt Brace Jovanovich, 1992.
McDermott, G. *Flecha Al Sol (Arrow To The Sun).* Viking Press, 1974.
Tan, A. *La dama de la luna (Moon Lady).* Lectorum, 1992.

Technology

Bill Walker Productions. *What Is The Brightest Star?* Available from Cornet/MTI Film & Video, (800) 777-8100. film, video, and videodisc
Broderbund. *Discover and Where In Space Is Carmen San Diego.* Available from Broderbund, (415) 382-4530. software
Cornet. *Our Sun and Its Planets.* Available from Cornet/MTI Film & Video, (800) 777-8100. film and video
Disney Educational Productions. *Johnson Space Center.* Available from Cornet/MTI Film & Video, (800) 777-8100. film and video
Hopkins Technology. *Amazing Universe.* Available from Learning Services, (800) 877-9378. cd-rom
LCA. *Beyond the Stars: A Space Star.* Available from Cornet/MTI Film & Video, (800) 777-8100. film and video
Learningways, Inc. *Learn About The Night Sky.* Available from Sunburst, (800) 321-7511. software